Agroecology in Action

Food, Health, and the Environment
Series editor: Robert Gottlieb, Henry R. Luce Professor of Urban and Environmental Policy, Occidental College

Keith Douglass Warner, *Agroecology in Action: Extending Alternative Agriculture through Social Networks*

Agroecology in Action

Extending Alternative Agriculture through Social Networks

Keith Douglass Warner

The MIT Press
Cambridge, Massachusetts
London, England

For information on quantity discounts, email special_sales@mitpress.mit.edu.

Set in Sabon by The MIT Press. Printed and bound in the United States of America.

Library of Congress Cataloging-in-Publication Data

Warner, Keith.
Agroecology in action : social networks extending alternative agriculture / Keith Douglass Warner.
    p. cm. — (Food, health, and the environment)
Includes bibliographical references.
ISBN-13: 978-0-262-23252-4 (alk. paper)
ISBN-10: 0-262-23252-9 (alk. paper)
ISBN-13: 978-0-262-73180-5 (pbk. : alk. paper)
ISBN-10: 0-262-73180-0 (pbk. : alk. paper)
1. Agricultural ecology—United States. 2. Agricultural innovations—United States. 3. Alternative agriculture—United States. 4. Agricultural ecology. 5. Agricultural innovations. 6. Alternative agriculture. I. Title.

S441.W28 2007
577.5'50973—dc22

2006046712

10   9   8   7   6   5   4   3   2   1

This book is dedicated to Bill Friedland, a most inspiring scholar.

Contents

Series Foreword   ix

Foreword by Frederick Kirschenmann   xi

Acknowedgments   xv

Introduction: Re-Thinking the Ecology of Industrial Agriculture   1

1 Rachel's Dream: Agricultural Policy and Science in the
  Public Interest   9

2 Agroecology in America: An Integrated System of Science and
  Farming   35

3 Cultivating the Agroecological Partnership Model   59

4 The Partners   89

5 The Practices   125

6 Agroecological Networks in Action   163

7 Circulating Agroecology   197

8 Public Mobilization   217

Notes   233

References   257

Index   271

Series Foreword

I am pleased to present the first book in the Food, Health, and the Environment series. This series explores the global and local dimensions of food systems and examines issues of access, justice, and environmental and community well-being. It will include books that focus on the way food is grown, processed, manufactured, distributed, sold, and consumed. Among the matters addressed are what foods are available to communities and individuals, how those foods are obtained, and what health and environmental factors are embedded in food-system choices and outcomes. The series focuses not only on food security and well-being but also on regional, state, national, and international policy decisions and economic and cultural forces. Food, Health, and the Environment books provide a window into the public debates, theoretical considerations, and multidisciplinary perspectives that have made food systems and their connections to health and environment important subjects of study.

Robert Gottlieb, Occidental College
Series editor

Foreword

As conceived by the Smith-Lever Act, the role of the Extension Service was to "diffuse" useful and practical information to people engaged in agriculture and home economics. It was assumed that new technologies, developed by researchers at land-grant universities, could bring the industrial revolution to America's farms and help the farming community interface more effectively with the evolving urban economy. The Smith-Lever Act, accordingly, defined the Extension Service as the "technology transfer" link between the research community and farmers. This agricultural paradigm defined researchers as active producers of information and farmers as the passive recipients.

With the advent of industrial agriculture organization—complete with input dealers, private agricultural management field specialists, and electronic information systems—it can be argued that the Extension Service's "technology transfer" role has become obsolete.

But agriculture is changing in ways that may make the role of the Extension Service more important in the future. This emerging agriculture is based more on ecological science than on technological innovations. Such agroecological initiatives assert that many current technologies based on fossil energy can be replaced with proper interactions between crops, livestock, and other organisms. And this new approach brings with it new challenges and opportunities, and a critical need for a new kind of Extension Service. It is this re-invention of the Extension Service that makes Keith Warner's *Agroecology in Action* such an important work.

Industrial agriculture was based on a set of core assumptions that largely held true for the past half century. Among those assumptions were the following:

Production efficiencies could best be achieved through specialization, simplification, and concentration.

Technological innovation would be able to overcome all production challenges.

Control management was effective.

Natural resource depletion could always be overcome, and sinks in nature would always be adequate to absorb the wastes.

Therapeutic intervention was the most effective strategy for controlling undesirable events.

Cheap energy would always be available.

As we enter the twenty-first century, all these assumptions are being challenged. Oil, natural gas, and irrigation water, which provide most of the inputs necessary for the continued success of industrial agriculture, are being depleted. The advent of climate change promises to bring us more unstable, volatile climate scenarios, putting highly specialized, monoculture farming systems—which require relatively stable climate conditions—at a comparative disadvantage. And, as the United Nations' 2005 Millennium Ecosystem Assessment Synthesis Report reveals, the ecosystem changes inflamed by our approach to acquiring food, water, timber, fiber, and fuel over the past half century has polluted or over-exploited two-thirds of the ecological systems on which life depends; furthermore, we have reduced both species and genetic diversity, which has dramatically reduced the resilience of ecosystems.

Meanwhile, pest-management specialists have come to recognize that "therapeutic intervention"—intervening in a system with an outside counterforce—as the sole solution to solving pest problems, at best, provides only short-term relief. Inevitably, in the long run, the attempted solution always becomes the problem as pests become more resilient and pervasive (source: W. J. Lewis, J. C. van Lenteren, S. C. Phatak, and J. H. Tumlinson III, "A Total System Approach to Sustainable Pest Management," *Proceedings of the National Academy of Sciences* 94, November 1997: 12243–12248). In addition, studies in ecology and evolutionary biology are rapidly teaching us that living systems, like farms, do not readily lend themselves to control management.

Consequently, farms of the future are likely to operate on a set of principles that differ significantly from the core assumptions that guided industrial agriculture. Farms of the future will likely have to be energy conserving, feature both biological and genetic diversity, be largely self-regulating and self-renewing, be knowledge intensive rather than energy intensive, operate on biological synergies, employ adaptive management strategies, practice ecological restoration, and achieve optimum productivity through multi-product, synergistic production systems that feature nutrient density, rather than monocultures that feature maximum yields. In other words, farms of the future are likely to operate on ecological principles.

While much research still needs to be done before we fully understand how ecologically informed farming operations can be managed in an optimally productive manner, the "greatest obstacle to ecologically informed alternative practices has not been a shortage of ideas, but more the dearth of practical educational initiatives—also known as 'extension'—to help producers learn about them," as Warner suggests.

Wendell Berry reminds us that "if agriculture is to remain productive, it must preserve the land, and the fertility and ecological health of the land; the land, that is, must be used *well*. A further requirement, therefore, is that if the land is to be used well, the people who use it must know it well, must be highly motivated to use it well, must know how to use it well, must have time to use it well, and must be able to afford to use it well." In other words, productive ecologically informed farming must acknowledge "the sources of production in nature and in human culture." (Berry, *What Are People For?* North Point Press, 1990, pp. 206–207)

Since such natural and cultural elements are essential requirements for successful ecologically informed farming, information cannot be reduced to transferring technologies. In this new world of agriculture, the Extension Service must be collaborative, must engage the farmers and the communities in which the farms exist, and must anticipate developing the kind of local knowledge, local economies and local culture that are appropriate to the local ecology in which the farms exist.

The collaborative arrangements which Warner's study identifies and describes can, therefore, serve as models for the new communities of practice that will need to emerge throughout the land if we are to meet the agricultural challenges that are rapidly approaching. And as the examples in this book demonstrate, a transformed Extension Service can play an important role in developing these new learning communities of the future. And this transformation on the farm and in the Extension Service can be further expedited by introducing new policies to facilitate such transformation, as Warner suggests near the end of this book.

Frederick Kirschenmann
Distinguished Fellow and former Director
Leopold Center for Sustainable Agriculture

Acknowledgments

With gratitude I acknowledge financial support from the National Science Foundation (award BCS-0302393), the Biologically Integrated Farming Systems Work Group under the leadership of Jenny Broome and Marco Barzman at the UC Sustainable Agriculture Research and Education Program, the UC Santa Cruz Department of Environmental Studies, and the UC Santa Cruz Center for Agroecology and Sustainable Food Systems.

Sean Swezey provided much support and encouragement as a member of my dissertation committee and as director of the UC Sustainable Agriculture Research and Education Program during the period of my research. Several partnership leaders were exceptionally generous in explaining to me their efforts: Cliff Ohmart, Marcia Gibbs, Pat Weddle, and Sean Swezey substantially shaped my understanding of partnership dynamics. Gary Obenauf, Fred Thomas, Kris O'Connor, Nick Frey, Mark Looker, Joe Grant, Mark Looker, Karl Arne, Chris Feise, and Bev Ransom all went beyond the call of duty in helping me.

This work is much stronger as a result of the input, suggestions and support of the UC Santa Cruz Agrofood Studies Research Group. Dustin Mulvaney provided helpful cartography, and Jill Harrison constructive feedback on my ideas. Daniel Press taught me much about the interactions between policy, politics, and society. Bill Friedland's decades of investigation into the social power of agricultural science has been an inspiration to me and my scholarship. Margaret FitzSimmons is an absolutely wonderful academic advisor, and I will be eternally grateful to her for all her instruction, mentoring, and friendship. My parents and family were unflagging in their enthusiasm, encouragement, and assistance. Most of all, I thank my Franciscan brothers for loving, encouraging, and supporting me during this journey of exploration.

Agroecology in Action

Introduction: Re-Thinking the Ecology of Industrial Agriculture

Science consists not in the accumulation of knowledge, but in the creation of fresh modes of perception.
—David Bohm (1993)

Rachel Carson's 1962 book *Silent Spring* upset American thinking about technology and the environment. Carson challenged society to think critically about the relationships between agriculture, science, and nature. Agriculture serves as the fundamental metabolic relationship binding nature and society, and our agro-environmental problems are symptoms of the underlying ruptures between them.[1] Carson exposed the serious harm that the agrochemical production paradigm visited upon human health and the natural world. Over time, agriculturalists and their allied industries have generally responded in one of two ways: by dismissing these problems, or by learning their way out of them.

This book describes how some farmers, scientists, agricultural organizations, and public agencies have developed innovative, ecologically informed techniques and new models of social learning to reduce reliance on agrochemicals, and thus put agroecology into action. It describes the technical scope and geographic range of ecologically informed strategies and practices in American agriculture, analyzing in detail a set of specific agroecological partnerships in California. This book explains how agroecology has become the primary approach to addressing the environmental problems of agriculture, although its lessons have implications for any initiative to deploy ecology for environmental problem solving.

Silent Spring was the first major book to raise serious questions about American industrial agriculture. Carson brought it to public attention that the agriculture on which we depend for sustenance was poisoning our environment and our very selves. Indiscriminate use of chemical technologies risked our planet and future generations. Carson made pollution visible, but also the inescapable ecological interdependence of society and nature. Her work led to the creation of the US Environmental Protection Agency, the banning of the most hazardous pesticides, and more regulation on the use of many others. Less well known are the innovations in agriculture her work inspired.

Four decades after the publication of *Silent Spring*, American society continues to wrestle with agricultural pollution. US agriculture has doubled its use of pesticides.[2] Agriculture is the leading cause of non-point-source water pollution in the United States.[3] Surface, ground, and coastal waters are seriously impaired, not just by pesticides, but also by nitrogen fertilizers. Carson never anticipated the scale and impact of nutrient pollution.[4] Farmworkers and rural communities continue to suffer from pesticide drift.[5] The indiscriminate use of agrochemicals threatens the future of Earth's biological diversity.[6] Agriculture has extensive negative environmental impacts, and these directly threaten rural communities and the health of American society. In recent years, scientists have begun describing the ecological services provided to society by agricultural landscapes, such as soil fertility, nutrient conservation, water quality, and flood control. Sustaining these kinds of ecological benefits from agricultural landscapes is essential to healthy societies now and will be in the future.

Carson sparked the modern environmental movement by making normative claims on society, inspired by an ecological vision. The scientific discipline of ecology assumed much greater social prominence as a result of her work and subsequent political advocacy. In the final chapter of *Silent Spring*, Carson described "the other road" of ecologically informed pest-management alternatives. There was, she wrote, an "extraordinary array of alternatives" based on ecological science. "Much of the necessary knowledge [of alternatives] is now available but we do not use it."[7] Since *Silent Spring*, ecology has proposed many more environmentally responsible practices, but they lack the immediate economic appeal of most other technological innovations. Environmental critics of agriculture (including the National Research Council) have repeated Carson's refrain, identifying alternatives and decrying their lack of adoption.[8] The greatest obstacle to ecologically informed alternative practices has not been a shortage of ideas; it has been the dearth of practical educational initiatives—also known as "extension"—to help producers learn about them.

Agroecological strategies and practices require more sophisticated knowledge, demand specialized labor, and in some cases may entail more economic risks for farmers, yet they hold out the greatest hope for reducing agriculture's environmental impacts. Despite the potential social benefits they offer, American agriculture's publicly funded scientific institutions have rarely embraced them, and when they have it has often been under duress. Since the 1980s, agroecological strategies have first been requested by farmers or farmers' organizations. Publicly funded agricultural science institutions, the land-grant universities and the Cooperative Extension Service, continue to operate out of a "Transfer of Technology" paradigm, delivering technologies to "end users," despite incontrovertible evidence that it is these technologies, chiefly agrochemical fertilizers and pesticides, that are environmentally problematic.

The Search for Alternatives

Many are the stories of agriculture's environmental problems, but this book is about initiatives to create solutions. It describes how agroecology has emerged as the primary conceptual framework for addressing modern agriculture's economic and environmental crises. Agroecology can be effectively put into action only when networks of farmers and scientists learn together about the local ecological conditions. Agroecology cannot be "transferred" in the way that a chemical or a mechanical technology can; it must be facilitated by social learning, which I define as participation by diverse stakeholders as a group in experiential research and knowledge exchange to enhance common resource protection.[9] Agroecological initiatives require a collaborative network to facilitate this social learning.

Since agriculture manipulates for human benefit the interactions of plants, animals, and the resources they need, ecological relationships are always present in farming, even when modern industrial technologies dramatically alter them. Farmers and scientists are now deploying agroecology for economic and environmental problem solving. Many authors are beginning to use the term "agroecology," but no previous work addresses the social relations that are necessary to support the extension of agroecology and its practical application.[10]

Using social learning to put agroecology into action has become the chief strategy for extending more "sustainable" alternatives within conventional agriculture. The social learning required to put agroecology into action relies more on the exchange of knowledge than on static expert knowledge or the delivery of technology. Producers, growers, farmers, their organizations, and scientists must be able to work collaboratively to share the various kinds of knowledge they bring, whether derived from the farm, the laboratory, or the marketplace. Those who manage farms must develop new skills of observation and decision making, but expert scientists must recognize that such skills are grounded in practical, local knowledge.[11] These extension activities have been much more successful with the help of a farmers' organization, one committed to farmers' needs and leadership, but capable of scientific competence and projecting a positive vision for agriculture. As this book demonstrates, in many cases growers and farmers have developed agroecological strategies and practices before agricultural scientists, and in some cases, despite the resistance of publicly funded institutions. In America, agroecological practice has usually led theory, and varied kinds of farmers' organizations have assumed new roles in facilitating the exchange of this knowledge through partnerships.

This book describes and analyzes the chief agroecological initiatives— their origins, organizations, networks, practices, and successes. It describes initiatives from every major region of the United States, with special emphasis on original field work in California. Surveying the most prominent agroecological initiatives, it explores the operational meaning of agroecology in the United States. In general, agroecological initiatives have been more successful in states with sustainable agriculture programs hosted by land-grant universities, although these programs have had uneasy relationships with their parent agricultural research institutions.

California's diverse specialty-crop agriculture has given rise to 32 partnerships in 16 different cropping systems. Analysis of these systems suggests important lessons for agroecological practice with relevance across all forms of agriculture in the industrialized world. The diversity of specialty crops and their emphasis on production for distant markets has led to farmers' being described as "growers." These California initiatives use what they call the "agricultural partnership model" to guide their extension activities, an approach that differs in important ways from conventional technology transfer. I restrict my use of the term "agroecological partnerships" to initiatives using alternative extension practices at a field scale, whether in California or elsewhere. These 32 partnerships form a coherent phenomenon, facilitating comparative analysis of the practice of agroecology in a specific regional context. This study carries important programmatic implications for any effort to help industrialized agriculture protect environmental resources, but it also raises questions about the character and practice of agricultural science. To interpret these efforts and controversies, I will draw from the field of Science and Technology Studies, and in particular from the conceptual framework of Bruno Latour.

Ultimately, agroecological initiatives seek to engage the broader society in support of this new approach to farming. Without supportive scientific and economic policies, agroecology will remain a subordinate, marginal practice. Taken as a whole, this study addresses critical issues at the nexus of food, environment, and health by describing how public policies shape the social context in which agroecological partnerships try to improve environmental management.

Organization of the Book

Chapter 1 opens with three vignettes tracing the secondary impacts of *Silent Spring* on the development of agroecological solutions to environmental and production problems in pear, winegrape, and almond farming systems. These provide insight into the diverse ways that growers, applied and research scientists, and growers' organizations have created networks to solve interlocking environmental and production problems. This chapter then recounts how Rachel Carson's vision affected agricultural regulation and science. She initiated a public outcry about pollution that led to strengthened pesticide regulations, although the US government has never created a comprehensive framework to address agriculture's environmental impacts. Agroecological initiatives have emerged to meet air and water pollution prevention goals that cannot be achieved through regulatory enforcement, in large part because of the diffuse character and ecological contexts of agriculture. Carson critiqued the contemporary practice of science as well, and agroecology has emerged as the primary scientific paradigm to guide new approaches to farming. This chapter also introduces Latour's circulatory model of science, which will serve as an interpretive tool for understanding the scientific controversies underlying research and extension in agroecology.

Chapter 2 narrates the development of two pioneer agroecological initiatives in Wisconsin and Iowa. Networks of intentional rotational graziers and a partnership between the Practical Farmers of Iowa and the Leopold Center at Iowa State University developed new models of farmer-scientist relationships in the American Midwest during the 1980s. Land-grant universities in the Midwestern states were modestly responsive to the expressed desire for new, more sustainable forms of production, and they expressed this through the creation of new programs. Constructively engaging scientific institutions has been a fundamental challenge for agroecological initiatives, because conventional agricultural science has deployed reductionist logic and has resisted taking a systems approach. This chapter introduces the land-grant universities and describes the controversies surrounding the operational understanding of their products, clients, and extension processes. Agroecological initiatives have emerged on the margins of these institutions but nevertheless need scientific expertise in order to succeed.

Chapter 3 describes and analyzes the evolution of the agroecological partnership model in California. Although the original almond-production techniques were developed by a farmer, the path-breaking BIOS almond partnership emerged from his interactions with an extensionist, a research scientist, and a staff member from a non-governmental organization. Acting together, these pioneers parlayed this partnership into a model for other commodities and for new granting programs. This has come to be known as the "agricultural partnership model." It is a loosely held set of assumptions about how various participants can work together to solve problems. Factors both internal and external to agriculture favored the expansion of this model since the mid 1990s. This chapter presents data on the progress three commodities have made in reducing pesticide use and improving their overall environmental performance. It also surveys national patterns of the partnership phenomenon.

Opening with a narrative of how a group of Washington State farmers worked together and partnered with others to create more sustainable forms of farming, chapter 4 describes the partners and their motives for association. Partnerships require active participation by farmers, scientists of all kinds, and agricultural organizations. This chapter reports on the characteristics of participating farmers and how they interpret the value of partnership activities. Agricultural scientists are configured in a hierarchy according to their primary type of scientific activity and economic sponsor, yet partnerships require the participation of diverse types of scientists, and successful partnership leaders have found ways to provide incentives for all of them to contribute. Partnerships mark the entrance of agricultural organizations into extension activities, and they do so as intermediaries between farmers and the public, as represented by markets and regulatory agencies concerned about pesticides and pollution.

New technologies and management techniques have shaped the ecology of farming systems. The narrative that opens chapter 5 illustrates how they pose challenges yet present opportunities for using agroecological strategies in almond farming systems. Partnerships use agroecological principles to guide the development and deployment of specific techniques to manage crops, animals, pests, nutrients, soil, and water; some partnerships capture synergistic benefits by managing these components of farming systems. This chapter presents an analysis of the five chief strategies of California's agroecological partnerships.

Using the scientific support necessary to implement codling moth pheromone mating disruption in pears as a case study, chapter 6 describes the social networks shaping partnership development. Successful partnerships have not created new relationships so much as intensified existing ones, weaving disparate relational threads into a common whole. The networks they constitute facilitate the generation and exchange of knowledge to better manage crop and non-crop organisms—and their ecological relationships—in farming systems. This chapter describes how strategic choices shape different configurations of networks, and how partnerships shape the development trajectory of agricultural research and extension.

Chapter 7 presents a model of what agroecological partnerships must have to successfully impact a commodity-production system. Its opening narrative explores the full range of initiatives one California winegrape partnership has undertaken. It describes the essential components of a successful partnership: human participants, scientific knowledge, dynamic social networks, and supportive off-farm institutions. It illustrates the value of Latour's model for understanding the scope of partnership activities.

Chapter 8 looks at the future of agroecology. Opening with a narrative about a potato partnership in Wisconsin, it shows how partnerships must be able to mobilize the larger public. The chapter draws conclusions about the significance of agroecological partnerships as means of addressing environmental problems, summarizes the principles that have made partnerships successful, and suggests what additional work must be done to fulfill the alternative, ecological vision of *Silent Spring*. Ultimately, ecological knowledge from the farm must flow outward to engage the public, and this chapter identifies the kinds of agricultural scientific and economic policies that must emerge to support them. Without such an integrated effort, American agriculture faces a dubious future.

Each chapter begins with a narrative of how partnership participants have come together to realize the possibility of alternative agricultural production practices, before turning to a more formal social science analysis. Michael Pollan's book *The Botany of Desire: A Plant's-Eye View of the World* inspired this descriptive approach. The narratives convey the dynamic interaction between plants, animals, and people. They provide an entrée into the socio-ecological world of farming. Collectively they describe an extensive cast of characters, but that is itself essential to this story. There is no one law, program or scientific expert driving these changes, and different partnerships follow their own paths to distinct ends. These networks are largely self-created and markedly less hierarchical than conventional extension. Three years of field interviews, primarily in California, provided information about these partnerships. These consisted of more than 135 personal interviews, 13 focus groups with 84 persons, and participant observation at 32 meetings. All quotes are from these interviews unless indicated otherwise. Additional source materials for the narratives and the social science analysis are documented thoroughly in the methodological appendix of my dissertation, upon which this book is based.[12]

1

Rachel's Dream: Agricultural Policy and Science in the Public Interest

Optimizing Eco-Rational Technologies

Doug Hemly's great-grandfather planted pear trees on Randall Island, about 15 miles south of Sacramento shortly after the California Gold Rush, making this the longest continually cultivated perennial crop production region in the state. The codling moth (*Cydia pomonella*) had plagued the fruit grown by his great-grandfather, his grandfather, and his father, so Doug had become accustomed to pesticides, although he had never really liked using them. During the 1970s Hemly and his father had actively cooperated with University of California (UC) researchers to develop ecologically informed strategies for using pesticides, timing their application carefully and using the least amount possible. This approach controlled the pest populations, but relied to the greatest possible degree on the beneficial action of predatory and parasitic insects. With the help of UC scientists, Pat Weddle had developed protocols for carefully tracking the peaks and valleys of insect population dynamics following the principles of Integrated Pest Management (IPM). Weddle, a professional pest-control advisor, advised Hemly about pest management for years. Their approach worked well until the bugs got uppity.[1]

Pesticide resistance develops within a population when repeated applications cull its most susceptible members, allowing only the fittest to survive and reproduce. Synthetic pesticides intervene in the ebb and flow of insect populations. This technology mimics the forces shaping natural selection. The codling moth is a particularly robust pest, having demonstrated resistance to lead arsenate in the 1920s and to DDT in the 1950s, in the latter case after fewer than ten seasons. After World War II, the chemical industries adapted organophosphates, originally designed as

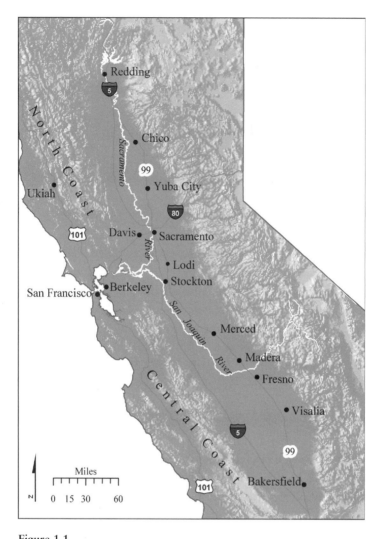

Figure 1.1
California's Central Valley.

weapons of war, to use against insects. Scientists and growers discovered they could use one organophosphate—azinphosmethyl (Guthion)—with their pear IPM strategies. By spraying it conservatively and precisely, growers managed their codling moth populations until 1991, when it stopped working. They applied the maximum legal rate—four sprays of 3 pounds of Guthion per acre—and still failed to stop the pest from eating into Hemly's income. When azinphosmethyl lost its efficacy against the codling moth, Weddle knew they had a problem. He contacted Stephen Welter, a professor and researcher at the UC Berkeley entomology department specializing in plant-insect interactions and the management of insect populations in agroecosystems. Weddle himself had received an MS degree from that department, inspired by *Silent Spring* and attracted to the ecologically informed initiatives undertaken by its faculty. Welter and his graduate student John Dunley devised a new methodology to assess resistance. They confirmed that codling moth populations in the Sacramento River pear district were now effectively immune to this pesticide.[2]

Fortunately, some Australian scientists had developed a novel product that releases a chemical mimicking the sex pheromones female codling moth release to attract the male for mating. Isomate dispensers appear like plastic twist-ties, and over a period of months they release their 0.0028 fluid ounce of synthetically produced pheromone into the orchard air, frustrating mating. In theory, flooding the orchard with artificial pheromones could disrupt the pest's reproductive cycle, but no one had ever done this in a commercial orchard. Because mating disruption is not lethal to insects, it would have to be deployed fully and consistently throughout the orchard when the codling moth is in flight. A failure to completely blanket the orchard with pheromones could result in an economic disaster.

Weddle had experimented with an earlier pheromone product on one orchard block, but used the regular organophosphate application to kill any additional codling moths that flew in from adjacent orchards. This proved effective, but at twice the cost. Weddle had more than 20 years of applied IPM experience, but the technical skills this experiment required were daunting. Welter thought the product could work, but only if the entire contiguous block of pear trees were managed to disrupt codling moth mating. He agreed to work on the project, but only on the

Figure 1.2
Doug Hemly shows off one of the pheromone mating-disruption devices he uses to control codling moth in his pear orchard. The synthetic pheromones are released from the plastic "twist-tie," which is attached to the tree with a modified bread bag clip.

condition that Hemly secure full cooperation from all the adjacent growers. Hemly spoke to the four other growers in the Randall Island area, three of whom agreed readily and one reluctantly, and they developed strategies to put all their 760 acres into pheromone mating disruption.

The Randall Island Project was widely considered a success by 1994, and began receiving awards and national recognition. This project was one of the most successful field-scale biointensive IPM projects of the 1990s.[3] In one season, Randall Island growers reduced Guthion applications by more than 75 percent. In the first year Isomate cost $220 per acre plus labor to apply, but by the late 1990s the price of the product had dropped and the growers had figured out how to use only half as much by applying it more efficiently. At $50 per acre, all the major pear growers in Sacramento County were using it. Participants had hoped to completely eliminate the need for organophosphates, but had to settle for dramatically reducing them.

The Randall Island Project's success built on the first IPM research in pears. During the early 1970s, pear growers in the Sacramento region had used as many as 14 active ingredients, most of them hazardous. This attracted the attention of early pear IPM researchers, then funded by the US Department of Agriculture, which began funding IPM projects after the pesticide controversies stirred up by *Silent Spring*. Pat Weddle recalls a flood in the Sacramento River Delta during the early 1970s in an area where some of the IPM trials were underway:

> Where they were using the chemically intensive program, those orchards defoliated during the flood. Where they were using the IPM program, they did not defoliate. You could drive down the river and see it. The Farm Advisor would say, there's a chemically intensive block and there is an IPM block. . . . Everybody could tell. It was a real eye popper. It made a lasting impression on what you could do if you could somehow manage pesticides in a way that was less disruptive.

High pesticide use stimulated organophosphate resistance in codling moth populations, but the early IPM work provoked interest in alternative practices among a network of leading growers as well. According to Weddle, "the Delta pear growers seemed to be that progressive core of people that were having enough trouble and were paying enough attention to the new possibilities."

The Randall Island Project relied on professional IPM-oriented entomologists to develop field monitoring protocols to track the behavior of the insect pest. The extra work these practices entailed did require extra compensation. According to Pat Weddle, ecological agriculture is information intensive, site specific, and labor intensive to monitor. Hemly was a member of the California Pear Advisory Board, a grower-supported organization whose goals were to foster marketing and facilitate research, and he was able to get that organization to help offset early additional costs of the pheromone products. In 1992 (the year Hemly, Weddle, and Welter began testing pheromones), Jean-Marie Peltier, the board's executive director, initiated the Pear Pest Management Research Fund. Sponsored by growers and canners, it provided about $150,000 per year to subsidize pheromone products. Contributing canners recognized that reducing the environmental and public perception risks of high pesticide use added value to the fruit they bought from growers and merited a higher price from their customers. Although invisible to anyone documenting the on-farm development of pollution-prevention practices, this joint research fund was critical to the success of the Randall Island Project.

By the late 1990s, Randall Island Project growers were down to four pesticides: oil, pheromones, one organophosphate, and a miticide.[4] By reducing the use of pesticides to control the codling moth, they allowed beneficial insects—formerly killed by organophosphates— to control some other pests that had required treatment. They were able to fulfill the vision of IPM of creating the right conditions in their farming systems to take advantage of "bug vs. bug" strategies, better known as biological control. The rate of pheromone adoption at the end of the 1990s was quite steep because its economic advantages were undeniable. Pear growers in other regions of California noticed the success of the project and began demanding access to it. Records kept by the California Department of Pesticide Regulation show that the California pear industry reduced organophosphate use 75 percent between 1994 and 2002.[5]

An Agroecological Imagination in Almond Growing

When Glenn Anderson returned to his family's farm, he knew he wanted to do things differently. He had grown up in California's Central Valley, 100 miles south of Sacramento, but had developed a certain dis-ease— physically and emotionally—with the ecological and economic implications of modern dairy production. In the 1960s he had dropped out of farming and moved to Hawaii, where he had taken courses in tropical agriculture and Pacific island ecology. He had begun reading Rodale's *Organic Farming and Gardening Magazine*, and in class he had inquired if organic agriculture could be put to use. His instructors had said that simply wasn't possible, and that chemical farming was the wave of the future. During the 1970s, Anderson traveled around the Pacific Islands and concluded that organic agriculture was not only possible but necessary. When he took over the family farm near Merced in 1980, he was convinced he could find another way to farm. His older brother had found almonds to be profitable, so Anderson set out to plant an orchard. Glen wanted to farm organically, but he didn't know how and had never met a scientist or an extensionist (that is, a field educator) who could help him.

A few years after starting his orchard, Anderson went to the "Ecofarm" Conference, where he heard about organic farming principles in vegetable crop production from instructors based at the (then named) Agroecology Program at UC Santa Cruz. He said to himself "That will surely fit an almond orchard." He began experimenting with leguminous cover crops and enhancing the ecological activity in his orchard. He drew from his study of ecology and the general principles of organic farming to guide his orchard management, relying on a diversity of organisms (cover crops, beneficial insects) to provide fertility and pest control instead of chemical technologies. At the end of the 1980s, he compared financial records with his brother and they discovered that Glen's costs were lower and profits were greater than his brother's conventional, chemical-intensive operation. If this were true, why were so many almond growers using so many chemicals?

Together the brothers approached their local UC Cooperative Extension Farm Advisor (a publicly funded extensionist), Lonnie Hendricks, about conducting a whole farm comparison study. Anderson and Hendricks both knew the director of the Sustainable Agriculture Research and Education Program (SAREP, located near the UC Davis campus), who readily agreed to fund the study. Anderson recalls:

At first [Hendricks] wondered "What the heck are these guys bringing to me? They're a little bit, especially the younger brother, they're odd." But as we did that, I really threw myself into it. I was very, very interested and excited about the possibility of doing this. And after the first year, Lonnie began to bring people around to say "Just a minute. Take a look at this." And of course, his colleagues said "You've got to be out of your mind, Lonnie. You're comparing just two orchards. I mean, that's not sound science. Anything can happen in that circumstance. Where's your replication?"

Hendricks had grown up on a pear orchard and had witnessed the chemical revolution. His father had used traditional practices such as cover crops and mineral oils, but Hendricks also recalls his family wearing gas masks as a farmhand sprayed parathion out of a tractor's exhaust pipe. They watched birds die as a result, and his father grew skeptical of pesticides. As a new Farm Advisor in the early 1960s, Hendricks read *Silent Spring*, and he felt it "made sense," although the work did not appeal to many of his Cooperative Extension colleagues. He expressed his opinion when asked, but didn't want to rock the boat too much.

As Hendricks began to talk about the curious results of the Anderson brothers' study, SAREP scientist Bob Bugg and Rick Reed, the program officer for the Community Alliance with Family Farmers (CAFF), were looking for a working example of ecological farming. Bugg and Reed had concluded that to advance the missions of their respective organizations, they needed to do more than critique conventional chemical practices, they needed to promote viable alternatives to it. When Reed and Bugg visited the Anderson orchard, they recognized what they had been looking for, and formed a partnership with Anderson and Hendricks.

They did not set out to promote organic agriculture, but rather to help conventional growers re-imagine their orchards as ecological systems. They wanted to help any grower recognize the practical economic and environmental benefits that could gained by farming according to ecological principles. If the conditions in an orchard merited a pesticide, they encouraged a grower to use one, but only if the grower was truly sure it was both economically and ecologically justified. While visiting Glenn Anderson's orchard, Bugg named this project the Biologically Integrated Orchard System (BIOS). Many people have observed that this was a brilliant name because, in the words of Glenn Anderson, it "resonated really broadly for all the right reasons." Hendricks then sent out a letter to the 800 almond growers in Merced County, inviting them to further participate in this study with the Andersons. Eighty growers came to their first meeting.

BIOS promoted a holistic, farming systems perspective. Instead of substituting one or two sustainable practices for a harmful technology, BIOS promoted a system re-design based on agroecological principles. BIOS worked with growers to help them perceive the potential interactions between components of their farming system and the advantages of a whole system approach. BIOS practitioners mentored growers as they applied ecological principles to the specific conditions of their orchards.

BIOS differed from conventional Cooperative Extension programs and their practice of technology transfer. Facilitating the acquisition of agroecological knowledge requires a different approach to social relations than transferring technology. It requires a social learning or co-learning model, in which leaders facilitate the exchange of learning experiences of all participants in practical research. BIOS developed local leadership around management teams consisting of growers, professional entomologists, Farm Advisors, and research scientists. This became known as the "BIOS model" or the "agricultural partnership model" of extension.

The first years of the Merced BIOS partnership demonstrated that this approach was possible and could be profitable. It required more effort, but it could be a more satisfying way to farm. CAFF developed a plan to expand BIOS to other counties and commodities, and to advocate with state and university leaders for more programs like BIOS. CAFF staffers were able to stimulate the creation of a larger, Biologically Integrated Farming Systems program at SAREP. They developed a strategic partnership with the Almond Board of California to promote this ecologically informed approach to agriculture, lending credibility to this approach. Inspired in part by BIOS, the California Department of Pesticide Regulation funded a program to help commodity boards conduct their own partnership activities to reduce pesticide use.

California almond growers achieved what appears to be the greatest volume of voluntary organophosphate pesticide use reduction in US history. Winter dormant season organophosphate use fell from a high of almost 500,000 pounds in 1992 to just over 100,000 pounds in 2000. BIOS and the Almond Board of California played crucial roles in educating growers about alternatives. Annual variation in weather, the economic advantage of new "softer" pesticides, and regulatory pressures from pesticide environmental contamination were also important factors, but they do not by themselves fully explain the scale of this reduction. The extent of organophosphate reduction varies by region, but counties with BIOS programs had the greatest reductions.[6] Agroecological partnerships have demonstrated that commercial scale monoculture can be substantially informed by ecological principles, if the requisite social relations are cultivated.

Cultivating Quality

About the same time that Hemly and Weddle began to notice the worrisome signs of pesticide resistance, a group of winegrape growers in adjacent San Joaquin County began devising an even more ambitious partnership. Located just north of Stockton, the Lodi area is the second oldest commercial winegrape region in California, with some of the operations dating back five generations of continuous winegrape growing to the 1860s. A hundred years later, Napa winemakers created their reputation by pursuing quality and promoting it through aggressive marketing. In contrast, the Lodi region had an established reputation for growing inferior winegrapes used only for cheap wines, and growers here were reputed to suffer from an inferiority complex. As the Napa growers continued to pursue and expand the premium wine niches, other regions could move up to fill the mid-price range. Because of its geographic setting near the cool breezes of the Sacramento Delta, the Lodi region had the potential to produce much better wines than its neighbors further south in the San Joaquin Valley. To enter the quality wine market would require grafting over most of their vines to varietal grapes—an expensive gamble—but maintaining the status quo would result in a slow but inevitable economic decline.

Randy Lange and John Ledbetter led a small group who wanted to emulate some of Napa's success in the Lodi area, but at that time, the State of California was suffering from yet another budgetary paroxysm. They recognized that to achieve their goal, they would need scientific resources, and that they could not rely on their traditional sources, the UC Farm Advisors. UC simply was not investing in field-based research to help the practical needs of agriculture as they had in previous generations. Lange, Ledbetter, and their colleagues were going to have to take matters into their own hands. They were going to have to fund their own research and extension program to add value to their product by distinguishing the quality of their wine and the quality of their environmental stewardship. They built on a history of regional cooperative marketing dating back more than 100 years. In 1986, they petitioned the federal government to create the Lodi American Viticultural Area. In 1990, Lange and Ledbetter lobbied the state legislature to amend the laws governing commodity organizations—previously requiring statewide agreement among growers—so that local crush districts could become their own winegrape commissions.[7] Lange, Ledbetter, and 25 other growers each put up $5,000 of their own money to campaign for the creation of the Lodi Woodbridge Winegrape Commission. Their original impetus was to foster applied research, but a significant number asked for help with marketing as well, which proved to be a popular selling point during the run up to the election. The commission was established by a majority vote of the region's winegrape growers in 1991.

Ledbetter saw part of the reason for wanting to fund regionally focused research and outreach was because leading growers in the region had come to the conclusion that sustainable pest management was the wave of the future: "Farmers are consumers, too. We get our food from the same supermarkets everyone else does. We drink the same water and breathe the same air. We have children and we can read. That's all it takes to make me want to look for alternatives to pesticides."[8] The Lodi Woodbridge Winegrape Commission began its IPM program in 1992 to "identify and implement sustainable strategies that will reduce conventional pesticide and fertility inputs while maintaining or improving the grower's net income."[9] The commission's five-point plan included using cover crops, improving soil health, monitoring for pests, pulling leaves to improve grape quality and reduce the risk of grape disease, and using the least hazardous and disruptive means of pest control. This program assumed that implementation of these practices would require addressing all those who participated in shaping decisions about a farming system: the grower, the professional entomologist, and the winery. Pest-control consultants would need to focus on monitoring the entire insect complex, not just pests, and the commission worked with wineries to help them clearly define acceptable pest damage levels so that growers would spray only when necessary to protect the quality of the winegrape flavor and not for cosmetic reasons.[10]

In 1995, Cliff Ohmart took over leadership of the IPM initiative. Ohmart had received a PhD in entomology at UC Berkeley in 1978 from the same department where Pat Weddle had trained and Stephen Welter taught. He strengthened the IPM program in quantity and quality. Ohmart obtained a grant from SAREP to integrate the grower-to-grower outreach model first developed by BIOS into the commission's organizational structure and identity. For demonstration vineyards, Ohmart selected 43 growers who put up the original seed money to start the commission, and they owned or managed more than half of the district's vineyard acres.

Leading growers fully bought into this outreach model, and were able to take advantage of the existing social relations among the 650 growers in the district. During the late 1990s, growers already familiar with IPM approaches applied those techniques more consistently throughout their vineyards, and those new to IPM were exposed to repeated encouragement to monitor their vineyards and make pesticide decisions based on site-specific data and economic thresholds.

The commission is widely credited within the California winegrape industry as creating the most comprehensive working model of a regional, grower-supported initiative to promote sustainable practices. Ohmart facilitated a change in how growers approach their farming systems by taking a region of predominantly conventional growers and helping them transition to using "sustainability" as the primary criteria for evaluating their practices. Growers do not understand themselves primarily as stewards, but have incorporated stewardship into their collective identity. His leadership is broadly perceived to be integral to the commission's overall success. While the commission advertises the quality of Lodi winegrapes, Ohmart travels as an ambassador of sustainability, describing how the commission has worked with growers to improve viticultural quality and environmental quality. The entire district has profited from this initiative, and the same growers are now considering how to launch an eco-label to further add value and engage the wine-consuming public in their project.

The California winegrape industry has invested more in partnerships than any other commodity. Four regional winegrape partnerships have worked particularly close to help local growers recognize agroecological alternatives to pesticides. The Lodi partnership has been the longest running and most visible of these. This partnership inspired a statewide winegrape growers' organization and a winery trade organization to develop the Code of Sustainable Winegrowing Practices. This is a comprehensive, objective, and transparent guide to growing winegrapes and producing wine in a way that is "environmentally sound, economically feasible, and socially equitable." Its sponsors have conducted 75 workshops around California, and reached more than 900 winegrape growers managing 120,000 vineyard acres (more than 20 percent of the state total). The code represents the most sophisticated program to date by any commodity-specific group of growers to educate themselves, and to present themselves to the public as environmentally responsible. Yet the Lodi Woodbridge Winegrape Commission still has an advantage lacking among most other groups: it has invested years of effort to develop a social network that provides peer and technical support for managing vineyards in the most environmentally friendly way possible.

Figure 1.3
University of California IPM Advisor Lucia Varela (to the right of the table)
explains the finer points of insect identification to growers using microscopes at
a Sonoma County Grape Grower Association IPM field day. Farm Advisor
Rhonda Smith (far right) discusses pest management with a winegrape grower.

Putting Ecology into Action

Most of *Silent Spring* narrated the apocalyptic implications of hazardous, invisible, carcinogenic biocides, but Carson concluded it with a
story of hope. She described how a few entrepreneurial scientists developed ecologically informed pest-management strategies. This chapter
describes the impact Carson's work had on both political and scientific
institutions shaping conventional agriculture, and how they, in turn laid
the foundation for agroecological partnerships, such as the three above.
Throughout this work I use the word "institution" to mean both
patterns of social behavior as well as human organizations necessary to
support them.

Carson made pesticide pollution a political issue, and appealed to the public interest as justification for doing so. Individuals have a right to protection from poisons introduced by others into the environment. She made normative claims: Only those able to understand the ecological hazards of pesticides should be allowed to purchase and use them. Regulatory institutions should be independent of political influence and provide safeguards for the public. Government should fully support the development of safe, ecologically informed alternatives. These were the chief elements of her public testimony before Congress in 1962, and were reflected in the famed Environmental Defense Fund lawsuits. *Silent Spring* started the public debate that resulted in the creation of the US Environmental Protection Agency (USEPA).[11]

The USEPA was created with a whole cluster of environmental laws, but two of them stand out for the purposes of this story.[12] Congress passed the Federal Environmental Pesticides Control Act in 1972, which transferred responsibility for pesticide registration and regulation to the USEPA, in large part to address charges that the US Department of Agriculture suffered from a pro-pesticide bias.[13] Congress assigned the USEPA responsibility for pesticide evaluation and registration, but regulatory enforcement of pesticide use was still in the hands of the states, with the USEPA supervising. In the same year, Congress passed the Federal Water Pollution Control Act Amendments, better known as the Clean Water Act. This set the ambitious goal that all surface waters be "fishable and swimmable" by achieving "zero pollution discharge" from point sources. This act gave the USEPA responsibility for setting standards for toxins in surface water, and allowed citizens to bring lawsuits to compel enforcement.[14] In 1974, Congress passed the Safe Water Drinking Act to protect public health by regulating the nation's public drinking water supply. American cities and industry have made impressive progress in controlling point-source, or end-of-pipe, water pollution. Agricultural nutrients are now clearly the greatest source of non-point-source water pollution.

The United States has never had an explicit, systematic environmental policy for agriculture, despite its large and widespread impacts.[15] Congress has never shown much interest in regulating the environmental consequences of agriculture, in large part because it has found the persistent cultural myths about family farmers, and the national virtues they exemplify politically useful.[16] Congressional leaders have protected agriculture from environmental regulation, even as this sector is increasingly responding to the wishes of vertically integrated transnational corporations.[17] Congress has never given the USEPA a clear mandate to address agricultural pollution, even as deleterious effects on water quality have become incontrovertible.[18]

The agricultural practices shaping the quantity, media, and impacts of pollution are distinct by crop, region, and specific ecological context, frustrating consistent regulation and enforcement. The kind of agriculture (annual crops, perennial crops, animal husbandry) determines the kinds of pollution (nutrients, soil erosion, agrochemicals) most likely to leak out of the farming system into the broader landscape. Local soil, moisture, and geological conditions have a tremendous effect on the severity and scope of pollution, and operation-specific management practices can result in highly variable environmental impacts, even within the same kind of cropping or animal production system. Agriculture can pollute several environmental media: air, surface water, and groundwater quality, often simultaneously, and sometimes from different activities. Environmental impacts may be temporally or spatially distinct from the activities that cause them, as is the case with surface and groundwater pollution. Decisions about agricultural practices in the United States are made by 2 million farm operators. Each of these factors presents a significant obstacle for typical environmental regulatory strategies, confounding the regulatory uniformity required of an equitable process.[19]

From a national perspective, nutrients such as nitrogen and phosphorus create agriculture's most extensive environmental problems in streams and groundwater.[20] When nutrients borne by rivers reach estuarine and coastal waters, toxic algal blooms, marine mammal deaths, habitat destruction, and shellfish poisoning can result. "Problem areas occur on all coasts, including those of California, Florida, Louisiana, Maryland, Massachusetts, New York, North Carolina, Texas, and Washington, but problems are particularly severe along the mid-Atlantic coast and the Gulf of Mexico."[21]

The Farm Belt's regional discharge of massive amounts of nitrogen through the Mississippi River has created a hypoxic, or dead, zone covering up to 7,700 square miles in the Gulf of Mexico. On the Eastern

Seaboard, nutrients from farms and confined animal feeding operations flow into estuaries causing algal blooms and outbreaks of *Pfiesteria piscicida*, a dinoflagellate causing fish lesions and fish kills, and perhaps posing a threat to human health.[22]

Nitrates threaten human health when they contaminate groundwater drinking supplies because they can affect the blood's ability to carry oxygen, especially when an infant's digestive tract converts it to nitrite, causing "blue baby" syndrome. Rural water supplies are susceptible to nitrate contamination from agriculture, especially shallow wells.[23] For example, on Washington's Columbia Plateau, nitrate concentrations in about 20 percent of wells exceed the drinking water standard, and the highest rates were found where fertilizer use is greatest. Nitrate concentrations in shallow wells here are among the highest in the nation. Agricultural fertilizers are the leading cause of nitrate pollution, followed by discharges from cattle feedlots and food-processing facilities.[24]

Even as *Silent Spring* provoked new environmental policies, US agricultural pesticide use grew dramatically, reaching a billion pounds per year in 1976 and fluctuating around that level ever since.[25] California's specialty-crop agriculture has used a disproportionate amount of the nation's total pesticides, roughly 20–25 percent.[26] After DDT was banned, many growers compensated by switching to organophosphate pesticides. These insecticides do not bioaccumulate and threaten top predators as did DDT, but they are acutely toxic, and increasing reliance on organophosphates meant greater acute health risks to growers and farm workers.[27] California's San Joaquin Valley has the maximum concentration of many pesticides among all of the US watersheds studied by the US Geological Survey: pesticides were detected in 69 percent of the groundwater samples collected from the eastern San Joaquin Valley.[28] A companion report on the Sacramento River found that watershed to generally be in better shape, although it did find that agricultural streams here have some of the nation's highest concentrations of the insecticide diazanon.[29]

Twenty years after the USEPA's creation, attention within the agency began to focus on agriculture's environmental problems. The USEPA was initially created with a legislated shotgun marriage of existing media (air, water) and category (pesticides, solid waste) programs. William Reilly, USEPA Administrator under President George H. W. Bush, directed his

staff to undertake regional analyses of environmental problems to determine gaps they needed to address. Integrated analysis of environmental
indicators in several regions indicated that agriculture merited priority
attention, and the USEPA launched place-based agricultural initiatives.
Under Reilly's tenure, the USEPA also emphasized public sector/industry
partnerships to promote voluntary pollution prevention.[30] The agency
also began initiatives to work with other federal agencies with existing
authority to manage resources. This shifted attention from end-of-the-
pipe pollution management to the reduction or elimination of potential
pollutants, especially hazardous or toxic materials.[31] These administra-
tive initiatives nudged the USEPA to address agriculture's problems, but
staff knew they had to do so in a non-confrontational way.

During the Clinton administration, the US Department of Agriculture
set new goals for implementation of IPM, but the passage of the Food
Quality Protection Act in 1996 was the most important agricultural regulatory initiative of that decade. As chapter 3 narrates, this law created
new concerns in the agricultural community about the loss of chemical
technologies, but also new funding sources for agroecological partnerships. As chapters 3 and 4 explain, when USEPA staff encountered BIOS
and other partnerships, they discovered that by investing in them they
could achieve agency goals with carrots, not merely sticks.

Essentially all industrial farming operations pollute, and comprehensive enforcement is impossible. Laws such as the Clean Water Act and
Food Quality Protection Act provide a framework for environmental
regulatory agencies at the federal and state levels, but their limited
resources leave all but the most egregious environmental offenses unaddressed. Agriculture's widely distributed and independent decision
makers, managing varied farming systems in highly variable ecological
contexts frustrate regulatory enforcement models. For these reasons,
developing incentive systems of collaborative voluntary innovation hold
more promise.

The Emergence of Agroecology

Silent Spring challenged scientific institutions as well as public officials.
Carson was one of the first in the Cold War era to openly criticize publicly funded scientific institutions for disregarding the public good.

Although she did not use the word "ecology," she advocated an approach consistent with its principles. *Silent Spring* inspired many students of the 1960s and the 1970s—such as Pat Weddle—to put ecological concepts into action. Some of those students became scholars who developed applied sub-disciplines of ecology, such as agroecology and conservation biology, to redress environmental problems. Both of these sub-disciplines make implicit ethical claims on the organization of society according to ecological principles.

Agroecology prescribes agricultural and ecosystem management strategies based on the discipline of ecology, and, as Stephen Gliessman argues, it marks a convergence between the agricultural and ecological sciences.[32] Like conservation biology, agroecology is a value-laden science that proposes ecologically informed concepts for solving socio-environmental problems, meaning they both assume the merit of environmental resource conservation.[33] Agroecology asserts that a farm should be managed as a functional system, and that wise farming decisions must be guided by an understanding of the structure and function of natural ecosystems. It carries with it an explicit criticism of reductionism, or scientific perspectives that pay attention only to individual

Table 1.1
Elements of agroecology.

Agroecosystem processes (from Altieri 2002)	Practices promoted by California's agroecological partnerships
Organic matter accumulation and nutrient cycling	Cover crops, application of compost, manures, chipping tree prunings, crop residue management
Soil biological activity	Cover crops, application of compost, manures, crop residue management
Natural control mechanisms (disease suppression, biocontrol of insects, weed interference)	Removing ecologically disruptive agrochemicals from the farming system, bio-diversification to attract and retain beneficial insects, cover crops
Resource conservation and regeneration (soil, water, germplasm, etc.)	Protection of streams with buffer strips, efficient use of irrigation water (no attention paid to genetic resources)
General enhancement of agrobiodiversity and synergisms between components	Managing components of farming systems to capture synergistic benefits

components of systems. It also recognizes that ecosystem services upon which society depends flow from managed landscapes, and that evaluating an ecosystem on the basis of a single indicator cannot capture the ecosystem's full range of benefits.

Miguel Altieri proposed three chief characteristics of agroecology: a systems framework of analysis, a focus on both biophysical and socioeconomic constraints on production, and use of agroecosystem or region as a unit of analysis.[34] More recently, Altieri has described agroecology as optimizing agroecosystem processes, which correspond rather well with the practices promoted by California's agroecological partnerships (see table 1.1).[35] The term "agroecology" was created in Mexico by ecologists, agronomists, and ethno-botanists as an agricultural development framework in opposition to the Green Revolution.[36] In its original form, agroecology helped peasants improve their indigenous farming practices as an alternative to high input, chemical-intensive agriculture.

Agroecology in advanced industrialized countries necessarily draws more from ecological sciences than from traditional, indigenous ethnobotanical knowledge more typical of the developing world. Over the past 50 years, industrialized US agriculture has discarded the ecological wisdom inherent in farming systems that integrate animals and crops. Here growers must learn to manage nutrients in ecologically informed ways, and devise alternatives to essentially ecologically irrational technologies (e.g., organophosphates). An agroecological or farming systems perspective identifies nutrient loss and non-target pesticide impacts as consequences of leaky, poorly assembled, and brittle systems.[37] Coping with these as symptoms is essentially a holding action, and authentic solutions require rethinking the design and management of agroecosystems based on principles derived from the study of natural ecosystems.

Agroecology proposes developing strategies and practices informed by the science of ecology, adding biological diversity to farming systems, and managing their interactions for synergistic benefits for the farmer and society. Ideally, nutrients would no longer leak from the system because they would be circulated back to other organisms, such as manures serving as plant fertilizers. "Pest" populations would not explode to economically damaging levels because the niches they exploit would be occupied by a range of other organisms, or their populations would be regulated by predators or parasites. For example, instead

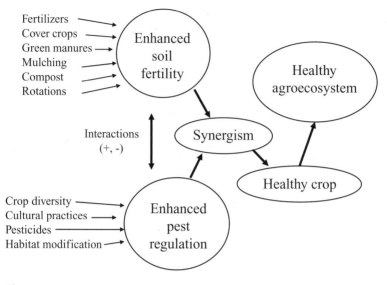

Fertilizers
Cover crops
Green manures
Mulching
Compost
Rotations

Enhanced soil fertility

Healthy agroecosystem

Interactions (+, -)

Synergism

Healthy crop

Crop diversity
Cultural practices
Pesticides
Habitat modification

Enhanced pest regulation

Figure 1.4
Interactions between components of a farming system (from Altieri 2002).

of repeatedly applying herbicides to control weeds on an orchard floor, a grower would plant cover crops, creating improved soil health, and reduce soil and water erosion. This approach to farming requires a different mode of perception, and the ability to recognize the potential benefits of managing components for superior outcomes for the entire system. Thus, agroecology's integrative, holistic approach exists in fundamental tension with reductionistic scientific practices and perspectives.

Agroecology is becoming a primary scientific paradigm to guide alternative agriculture, partially replacing the term "sustainable agriculture" within the academy. "Sustainability" is a compelling albeit slippery term, and continuing debates about and differing interpretations of its precise definition make it difficult to use without having to continually define it.[38] Many sustainable agriculture initiatives in US agriculture have in practice been narrowly focused, on single practices within farming systems. I prefer the definition of Patricia Allen et al.: sustainable agriculture is "one that equitably balances concerns about environmental soundness, economic viability, and social justice among all sectors of society."[39] In practice and program, however, collaborative progress toward this comprehensive and ambitious framework to sustainability has been difficult to achieve.[40]

One does not have use the term "ecology" to recognize ecological relationships and the benefits of manipulating them. Growers and their
consultants appear comfortable using products and practices described
by the stem "bio." Partnership leaders have assiduously avoided using
any terms with the stem "eco," even though they regularly take advantage of ecological relationships between organisms. This example is but
one of the discursive framing strategies used by leaders of these agroecological initiatives.[41]

For the purposes of this study, I consider the terms "holistic resource
management," "alternative agriculture," "biointensive IPM," and
"integrated farming systems" to fall under the broader umbrella of
agroecology. These will all be discussed in subsequent chapters. Organic
agriculture has also grown in recent years, although the federal organic
certification program has also led to contradictory outcomes.[42] In
its ideal form, organic agriculture is organized along agroecological
principles, although in practice this is rarely the case.[43] "Regenerative
agriculture" and "natural systems agriculture" are other terms similar to
"agroecology."

By framing the initiatives described by this book with the science of
agroecology, I hope to show how participants are using applied ecology
to undertake socio-environmental problem solving in agricultural production. Agroecology is concerned about the broader social and
economic context of farming, but attends primarily to questions of
agricultural production. Whereas authentic sustainable agriculture
encompasses all questions of food consumption, I have limited my investigation to agricultural production. Questions of social equity, economic
viability, and public policy lurk in the background of this book, and will
reappear in the final chapter to inform its conclusions.

Latour's Tool for Interpreting Scientific Controversy

To understand agroecology's critical stance toward the broader field of
the biological sciences, I will draw on Science and Technology Studies,
which now informs many studies of controversies within scientific
institutions. STS emerged to critique the overly idealized depictions of
scientists and their activities. Bruno Latour's 1987 book *Science in
Action: How to Follow Scientists and Engineers through Society* broke

open new ways of understanding the role of scientists in society. Latour and other STS scholars insist that science must be studied in action, as a variable, contingent, creative, and at times controversial cultural practice, not as plodding conformity to a static, trite formula to accumulate increasingly larger data collections. From an STS perspective, science is not so much the pursuit of "truth-to-be-revealed" as a social project weaving together knowledge work, organisms, technologies, theoretical arguments, and public representation. Latour insists that to truly understand science one has to follow scientists through society and observe them as (social) *actors*, in *action*, and this book derives its title from his work. He developed a methodological approach that analyzes scientists' engagement with the social world, not only discrete activities in a cloistered laboratory.[44]

STS investigates how actors use science knowledge to form networks and persuade other people, organisms, and technologies to collaborate.[45] Networks emerge as actors recognize that they are better able to feed and grow by associating with others, human and nonhuman, through a process Latour describes as "enrolling." Knowledge, therefore, serves as an incentive, piquing the interest of the scientist and others with the possibility of rewards. Latour refined his theory of hybrid social/scientific networks in *Pandora's Hope*, proposing a "circulatory system of science" (figure 1.5). According to Latour, "there are five types of activities that science studies needs to describe first if it seeks to understand in any sort of realistic way what a given scientific discipline is up to: instruments, colleagues, allies, public, and finally, what I will call *links* or *knots* so as to avoid the historical baggage that comes with the phrase 'conceptual content.'"[46]

For Latour, scientific knowledge is more powerful and more persuasive as it flows through society. He rejects the portrayal of science as an activity that is more real because it is "pure," isolated from contaminating social interests; instead, he proposes science as a beating heart at the center of a circulatory system of arteries and veins, pumping knowledge as though it were oxygen through tissue. Latour proposed this model as a tool to understand science generally, but I prefer its adaptation by Margaret FitzSimmons for understanding the circulation of ecology in society.[47] After introducing the dynamics of the five loops, I will deploy Latour's interpretive tool to analyze the controversies surrounding the work of Rachel Carson.

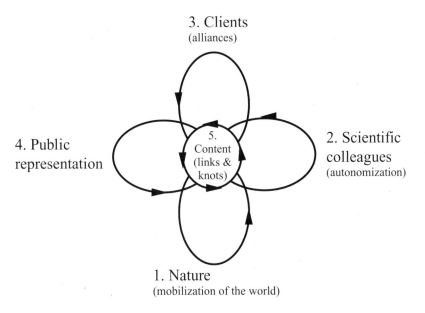

Figure 1.5
Latour's circulatory system of science.

In the first loop, which FitzSimmons titles "nature," humans investigate natural phenomena, which in turn shape human social behavior. Farmers have observed the natural world since the beginning of human civilization. More recently, scientists have joined them in the field, armed with the scientific method, statistics, and devices to make obscure phenomena visible. They discover how plants (crops, weeds, cover crops) and animals (livestock, insect pests and their natural enemies) behave in association in an environment of variable light, heat, water, and soils, and together they develop insights for improving agriculture. Knowledge circulates from the field to the university laboratory and then back out to the field. Scientists organize and categorize their discoveries in the world, and make them available for scholarly argument.

The second loop, which FitzSimmons titles "scientific colleagues," describes how scientists assemble facts and construct institutions devoted to knowledge. Scientists attempt to validate their discoveries by organizing knowledge into disciplines, sub-disciplines, specializations, and scientific communities that will amplify, feed, and validate further investigations. Peer approval is critical for the reproduction of science.

Without the imprimatur of colleagues, resources dry up, disciplines wither, and lines of scientific investigation fail.

Latour names the third loop "alliances," and FitzSimmons renames it "allies," but for my purposes it is best labeled "clients." This loop represents the patrons that scientists have always had. These are the people and groups that can put knowledge yielded by scientific discovery to practical use. Knowledge contained by the academy is feeble compared to that put into economic action. Agriculture may request specific help on problems from scientists, or scientists may need to persuade growers of the commercial potential of their research.

Latour labels the fourth loop "public representation," although he is quick to dissociate this from the stigma of "public relations." He insists that society ultimately has a say in the conduct of science, however distorted or stereotypical its understanding may be. The public is located just outside the edge of this loop. Pumping knowledge through it requires very different skills than laboratory investigations, but it too is science, just as much as is statistical analysis. This loop holds opportunities as well as dangers. Scientists may be celebrated for their innovation, or their credibility may suffer attack for making public knowledge from the field and laboratory, as Carson herself discovered. Scientists discount the importance of this loop to their peril, as evinced by public squeamishness about transgenic crops.

Scientific knowledge circulates by the pumping of the heart. Without "links and knots," or scientific content, there would be no other loops; they would die instantly. But the heart does not exist in isolation: it exists to distribute knowledge throughout the entire system. This conceptual core becomes stronger the faster it circulates knowledge through the other four loops, in turn strengthening the power of science within society.

Carson captured the public's attention by describing the ecological folly of indiscriminate agrochemical use. She critiqued the practice of scientists who thought of their work as exclusively serving their immediate, economic clients: actors in loops two and three were only concerned with each other's interests, and not ecological or social consequences. Carson diffused knowledge from loop 1 to loop 4, mobilizing the public to act on behalf of nature; in other words, she circulated knowledge from society to reduce the hazards of irrational pesticide use. For her efforts

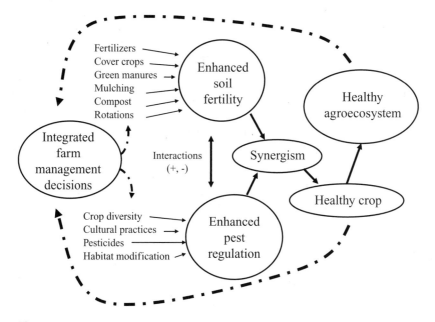

Figure 1.6
Interactions between components of a farming system and farm-management decisions (expanded from Altieri 2002).

she was savagely attacked by industry and elected officials, and the integrity of her scientific work was assailed. Carson was dismissed as an irrational, emotional woman, and her scientific work judged illegitimate. Her critics were wrong, of course, and her arguments about the public significance of the scientific consequences of irresponsible chemical use carried the day, in large part because she circulated her findings to the public, which recognized the interdependence of all these loops even when most agricultural scientists did not.

Using insights from Latour's understanding of science as practice, we can improve Altieri's conceptual model of agroecological interactions of systems components by adding a feedback loop, circulating knowledge back to the humans who make management decisions (figure 1.6). *Agroecology in Action* investigates how people have successfully circulated this knowledge back to farming decision makers for social and environmental benefits. In so doing, they collectively carry forward Rachel Carson's dream.

Conclusion

The three partnership vignettes that opened this chapter fulfilled Carson's dream of agricultural alternatives, inspired by an alternative vision of nature-society relations. They demonstrate how an ecologically informed conventional agriculture can do a better job protecting environmental resources. *Silent Spring* has had ripple effects through every institution shaping the agriculture/environment interface in the United States. Carson made agricultural pollution a political issue and laid out an alternative path, one informed by ecology. She challenged public officials and scientists to better safeguard the public interest. Agricultural pollution is particularly intractable, even though it now is the greatest source of non-point pollution. The diversity and variability of its origins and consequences have resisted regulatory and scientific initiatives to control it.

Legislative and regulatory responses to *Silent Spring* are more widely known than those of scientific institutions, but these too have been important. Scientists are now proposing agroecology and its integrated systems approach as the best framework for preventing pollution and protecting environmental resources while ensuring an economic return for growers. The scientific institutional obstacles facing this integrated farming system approach—and public interest strategies to surmount them—is the chief topic of the next chapter. Latour's model of science will serve as a guide for interpreting the controversies we will encounter.

2

Agroecology in America: An Integrated System of Science and Farming

Reversing the Logic

Mike Cannell recognized that his conventional dairy farming system just wasn't working any more. Costs were rising as milk prices fell, and he was working harder and harder only to make less. He had to re-think his approach to farming or get out. His friend Jim Brown had read about New Zealand's intentional rotational grazing strategies, and the discussion groups organized by farmers to exchange knowledge about them. They began to experiment with intentional rotational grazing individually, but consulted frequently with each other to try to understand its ecological dynamics. They recognized its multiple benefits, but realized they had a lot to learn. A network of peers like the New Zealand discussion groups would help them share their knowledge of grazing more efficiently, and it would help them learn more. In reality, they had no other choice as they started the Ocooch Grazers Network in 1993, because the University of Wisconsin (UW) offered these "grass radicals" no information on this technique. As Neva Hassanein documents in *Changing the Way America Farms: Knowledge and Community in the Sustainable Agriculture Movement*, these Wisconsin graziers developed new strategies to generate, exchange, and extend alternative scientific knowledge. Jim Brown and Mike Cannell realized that they were on to something when so many of their neighbors began calling them about their "dairy heresy."[1]

For 100 years the University of Wisconsin had invested in developing sophisticated science and associated technologies that inform conventional dairy production, which is now an industrial activity, intensive in capital, technology, and labor. It is predicated on the economic logic of breaking down the farming system into components, maximizing the productivity of these, and manufacturing the greatest commodity output. This has resulted in steady increases in the production of milk, but with many "unanticipated" consequences. The system provides non-stop, maximum flows of nutrients to the herd, but requires a significant investment in growing feedstuffs, and then delivering these to the cows, which in turn demands the dairy farmer to milk cows year round. The increases in corn and soy production used to feed cows, stimulated in part by federal subsidies, have had major, long-term environmental impacts. The nitrogen chemical fertilizers applied to cornfields were beginning to show up in Wisconsin groundwater, contaminating it and rendering it undrinkable. Confined cows produce about 50 pounds of manure and urine per day, which have to go somewhere.[2]

Nitrates from fertilizer and manure are of particular concern in southwest Wisconsin because of its sandy soils, which allow the movement of agrochemicals into groundwater. In addition, environmental researchers in Louisiana were beginning to document the impact of nitrogen from the Mississippi River watershed on marine life in the Gulf of Mexico. The socio-economic impacts of this production strategy are equally significant. Making the system economically profitable requires greater inputs of capital, machinery, and labor. Smaller family farms are lost when larger dairy farmers buy out their smaller, less competitive neighbors. The acceleration of inputs and production driven by these factors has been called a "technological treadmill."[3]

Intensive rotational grazing reverses the industrial logic of conventional dairy cow husbandry by re-designing the feeding system. Instead of plowing, planting, fertilizing, spraying herbicides and insecticides, harvesting, storing, mixing, and delivering feedstuffs to continuously confined animals, the grazier leads the cows to a pasture enclosed by an electric fence, and then rotates them to various paddocks depending on the relative availability of grass. Hassanein and Kloppenburg describe the system:

. . . monocultures are replaced by perennial polycultures. . . . Rotational grazing greatly reduces the need for specialized and expensive machinery, petroleum products, and agricultural chemicals. Manure handling is required much less frequently because cows only spend time in the barn during milking. Many graziers are trying to "go seasonal" by synchronizing the cows' lactation with pasture growth in the region. Seasonal milking can allow farm families to "dry off the herd," reduce winter confinement feeding costs, and take vacations when cows do not need to be milked. Farmers who have switched to intensive grazing management also report a reduction in labor requirements, improved animal health, greater per acre productivity, better profitability, and greater satisfaction from their work.[4]

With these kinds of benefits, why wouldn't everyone want to practice intensive rotational grazing?

In reality, this "new" approach demands new ways of perceiving and thinking, which require the management of new risks and the development of new approaches to farming. Graziers have to develop new skills of observation, watching the cows and the grass to make sure both are healthy, and they have to be prepared for seasonal changes specific to their pastures. Intensive rotational grazing demands flexibility, imagination, and careful attention to biological details (which practitioners call "the grass eye"). Graziers must be able to observe and interpret agroecological indicators of plant-animal interactions in their pastures. In doing so they are influenced by topography, soils, moisture, and previous management decisions. In the words of Jim Brown: "You have to look ahead and determine when to graze and that requires experience and observation. There aren't any rules or regulations. You gotta look at the pasture."[5] The spring flush requires special attention. At the beginning of the Wisconsin growing season, the lengthening days and warming temperatures stimulate the rapid growth of pasture plants, and the grazing animals have to be rotated through the several pastures quickly to prevent vegetation from overgrowth and waste. At the end of the season, in the fall, pasture growth is slow, and the animals have to be rotated through pastures fairly quickly so as to not damage the plants.

Intentional rotational grazing requires an agroecological knowledge system—what Cannell and Brown call a "knowledge base." Graziers developed local, participatory sustainable agriculture networks not because UW did not have the potential to help them, but because it was not doing so. For four decades, dairy research had investigated how to optimize the confined feeding system and the monocultures on which it depended. They had considered researching grazing, but had ruled it out because it is not as productive per acre, and thus they could not offer graziers knowledge about this farming system. As one UW dairy researcher confessed, he had nothing to offer them.

Cannell and Brown carried forward the historical tradition of farmers sharing knowledge with their neighbors that pre-dated the land-grant university extension system, but they were also inspired by the relatively new sustainable agriculture movement and its efforts to develop alternative sources of knowledge. Pioneering intentional rotation graziers decided to conduct their own research. They began by creating local, alternative networks of peers to meet their immediate knowledge needs. Hassanein traces the history of Wisconsin's sustainable agriculture networks back to a 1986 workshop in LaCrosse, sponsored by the Rodale Institute. This workshop described new research into organic and ecologically informed production techniques, and inspired several farmers to start their own local research efforts. They were critical of the UW agricultural research agenda, but they were not content to simply complain about it. They wanted to develop the alternatives themselves.

The Southwestern Wisconsin Farmers Research Network developed several novel strategies for facilitating farmers' participation in research and extension. It consulted farmers to identify the most helpful, needed, and practical research, the kind that would benefit farmers. The research would take place on real farms at commercial scales. The network cultivated relationships with the few, individual UW researchers open to working collaboratively with farmers. Network participants knew they could have greater impact if their research could boast of both "methodologically valid results" and practical outcomes for farmers. They organized regional field days to talk about their research. Prompted by skilled network staff, local newspapers provided good publicity. Participants in this network began to perceive that they were not merely exchanging knowledge about alternative farming systems, but that together they were learning to think in new ways. One grazier said: "For years, I had fought and borrowed [money] to bring feed to the cow. Then I figured out that I could bring the cow to the grass."[6] Other dairy farmers who converted to grass production subsequently reported that they began to question the very need for agrochemicals in their farming system, a truly subversive proposition in American agriculture.

Members of the network realized that they had an approach to helping other farmers develop more sustainable forms of agriculture, and they wanted to share this approach with others. With the help of a cluster of allied non-governmental rural organizations (especially the Wisconsin Rural Development Center), the network sought to institutionalize this approach to developing and extending alternative agricultural knowledge. In 1986, these groups coordinated a successful grassroots effort to establish a funding stream of state dollars at the Wisconsin Department of Agriculture, Trade, and Consumer Protection. The Sustainable Agriculture Program was one of the first in the country to provide a funding source for local groups to conduct their own research and extension activities. It lasted only 10 years, but it played a critical role in spreading the idea of local networks of farmers and graziers conducting their own research and outreach throughout the state. Mike Cannell had attended network field days and had been on the steering committee for the Wisconsin Rural Development Center. When he and Brown started the Ocooch Grazers Network in 1993, they were able to draw from successful strategies in organizing rural communities for positive social and environmental change from the center and the network. These sustainable agriculture networks were by no means the first initiatives of their kind. At the beginning of the twentieth century, public university scientists regularly contributed their expertise to regional farming groups. Multi-day farmers' institutes were held throughout rural Wisconsin. The success of these initiatives would later build support for a national system of agricultural science extension.

The Morrill Act of 1862 had launched the land-grant university system, creating what would become the world's largest agricultural science institution. With the scientific method, researchers addressed practical agricultural problems and increased yields, "making two blades of grass where only one had grown before." One unintended consequence of this arrangement was that agricultural scientists became "the experts" and farmers became "the clients," which conveyed special prestige and authority on scientists and their institutions.[7] These scientists helped some farmers become extremely successful using industrial strategies, but most smaller, "economically inefficient" farmers were eventually squeezed off their farms, and many producers who did remain became passive recipients of expert knowledge generated by universities.

In reality, farmers have conducted experimentation and exchanged knowledge since the dawn of agriculture. Across the Mississippi River from Wisconsin, Dick and Sharon Thompson began farming in 1950 near Ames, Iowa, after Dick received a master's degree in animal production from Iowa State University (ISU).[8] They farmed corn, soy, cattle, and hogs, but became increasingly dissatisfied with conventional approaches, and began experimenting with ridge-till strategies for growing corn and soy. Instead of cultivating the entire field, ridge-till farmers establish and maintain contours in their fields with ridges 6 to 8 inches high. Each spring, they till only the top few inches of the ridges for the seedbed. These ridges provide more exposure to sun and better drainage, which overcome any problems associated with crop residue. Instead of spraying the entire field with herbicides, ridge-till farmers apply thin strips over the ridge-top seedbed, and cultivate between ridges. Instead of spraying the entire field with chemical fertilizer, ridge-till farmers apply precise amounts when the emerging plants need it most. Instead of maintaining a barren field susceptible to wind and water erosion, ridge-till strategies use crop residues to hold soil in place. Research conducted by David Lighthall during the 1990s documented that ridge-till farming typically use only half the herbicides and three-fourths the nitrogen of conventional techniques.[9] From 1967 to 1983, the Thompsons purchased no fertilizer or herbicides.

In the early 1980s, farmers and farm advocates from Iowa banded together to address the crisis in agriculture that was sweeping across the upper Midwest. The collapse of commodity prices was undercutting the economic viability of farming, the negative environmental consequences of conventional farming practices were mounting, and the fabric of rural communities was unraveling with the demise of thousands of family farms. In 1985, Practical Farmers of Iowa (PFI) was founded to generate and exchange practical knowledge generated by farmers for farmers. Dick Thompson was elected as PFI's first president.[10] The creation of PFI formalized the way Thompson had been conducting research and extension on his own farm for years. Hundreds of farmers, rural advocates, teachers, and extensionists had visited his farm every summer throughout the 1980s to learn about his planting techniques, crop rotations, and re-incorporation of animals into in his farming system. PFI's founders were convinced that Thompson—and other farmers like him—had developed strategies that could address the crisis in farming, but they could not get the attention of ISU researchers. They decided to take matters in their own hands and organize on-farm research, using randomized and replicated plots that would yield statistically valid results. The few scientists at ISU who were favorable toward sustainable agriculture—chiefly Jerry DeWitt, the director of the Extension Service, and John Pesek, the chair of the agronomy department—were able to interest a few of their scientific colleagues in these findings.

Meanwhile, in the state capital of Des Moines, an unusual collaboration of public officials wanted to address a major environmental symptom of the farming crisis: groundwater pollution. In 1987, a bipartisan group of legislators was able to pass the Groundwater Protection Act, creating the Leopold Center for Sustainable Agriculture at ISU. This became the first sustainable agriculture program at a land-grant university in the farm belt. Its mission was to conduct research into the negative impacts of agricultural practices, to help develop alternative, sustainable practices, and to work with the university's extension system to educate the public about these findings. It was to be funded with state fees on pesticide and nitrogen fertilizer sales. It was bitterly opposed by the agrochemical industry, but the severity of the farm crisis created the right context for the bill's passage.[11]

Practical Farmers of Iowa and the Leopold Center at ISU have developed an extraordinarily close and fruitful partnership since the late 1980s. DeWitt enhanced the legitimacy of alternative agriculture at ISU. The Leopold Center provided PFI office space, and ISU provided PFI staff the status of university employees. The Leopold Center provided substantial funding to PFI farmers to conduct on-farm research and extension. As a public interest organization, PFI was able to draw attention to the good work of the Leopold Center, and it has fought passionately to defend its viability and budget. Inspired by this partnership, ISU has created graduate studies programs in sustainable agriculture. Partnerships like the Wisconsin graziers' networks and the PFI/Leopold Center do more than reverse the logic of industrial agriculture. They bridge the gap between universities and the practical needs of farmers, and they reverse the collapse of hope in rural communities. Before he retired, John Pesek chaired the National Research Council's Committee on the Role of Alternative Farming Methods in Modern Production Agriculture, which in 1989 produced the landmark report *Alternative Agriculture*.[12]

Back in Wisconsin, a coalition of farmer and environmental organizations, including the Wisconsin Rural Development Center and the

Southwest Wisconsin Farmer Research Network, worked with some UW faculty to propose an alternative way of integrating farmer and rural community interests with university research. After a two-year process, UW's College of Agriculture and Life Sciences established the Center for Integrated Agricultural Systems (CIAS) in 1989. This center collaborates with stakeholders to identify research priorities, conduct research, and extend this knowledge (frequently through networks). The coalition persuaded the legislature and the governor to fund CIAS as a permanent budget item.[13] In the early 1990s, dairy farmers knew more about intentional rotational grazing than did scientific researchers. One of CIAS's first projects was to convene graziers, extensionists, and scientists on equal footing to conduct research of practical value to farmers. As of 2005, more than 20 percent of Wisconsin dairy farmers practice some form of rotational grazing.[14]

The Institutional Ecology of American Agricultural Science

Silent Spring was published on the centenary of the Morrill Act, and it heralded the end of an era. For its first hundred years, the land-grant university (LGU) system, chartered by Congress to assist pioneer America with the practical science to build the country, was seen as an unqualified success. Indeed, up until World War II it was probably the largest single scientific institution in the world. Funded by public tax dollars, the LGU system produced an astonishing quantity of scientific knowledge and practical technologies. The agricultural production systems designed by LGU researchers, fueled by billions of dollars of federal crop subsidies, produce far more food than America could ever possibly consume, but at substantial social and environmental cost. Rachel Carson launched a line of questioning that continues to provoke debate about the role of LGUs in American society: Do they produce the kind of agricultural science the public really wants?

Agroecological initiatives in industrialized countries face two fundamental problems: persuading agricultural science institutions to provide expert, ecologically informed knowledge, but to do so in a way that facilitates the active engagement by farmers in learning about the particularity of their own farming system. The stories from Wisconsin and Iowa relate the challenges facing farmers and rural communities as they sought scientific assistance in developing alternative farming systems. The intentional rotational graziers and PFI expressed something akin to "buyer's remorse," because they had tried the dominant, industrial high production systems designed by LGU scientific experts and found their "unanticipated" consequences unacceptable.

The narratives in this book relate how farmers have set out to create alternatives to conventional, LGU designed farming systems. The farmers and their organizations were not anti-science, but instead demanded a different kind of science, a more holistic form, integrating social and environmental considerations. The clamor for alternative, agroecological farming systems grew out of multiple critiques of the LGUs' scientific products, their conceptualization of their clientele, and their approach to conducting extension.

Latour's model (figure 1.3 above) helps us interpret the socio-scientific dynamics underlying these conflicts. Latour created this model because all too many scientists and science policy administrators believe that science consists exclusively of the knowledge circulating between loops 2 and 5, scientific colleagues and scientific content. So long as some clients (in loop 3) are content with their products, they are generally unconscious or unconcerned about the social implications of their science work. The industrial grain and dairy production systems designed by LGU scientists were remarkably productive, and thus initially considered a success. The LGUs have sustained decades of criticism because their products and extension processes have generally failed to consider negative environmental (loop 1) and social (loop 4) impacts. The fruits of LGU work certainly have benefited some agricultural clients (in loop 3), generally helping large operations become more efficient, pushing smaller farmers out of farming, and furthering the economic concentration of agricultural production in the hands of a few. Critics drew attention to these consequences and asked the public (which was paying the bills, and located just outside loop 4) to support alternatives. When LGU scientists and administrators have admitted that their products and extension processes are in part responsible for these negative consequences, it has generally been under public pressure. The persistent controversies dogging the LGU system turn around questions of what kind of scientific knowledge is produced for whose benefit, and how that knowledge shapes both society and nature.

This chapter traces the origins, purpose, and importance of the LGU system, and recounts how those searching for agroecological solutions have tried to engage this institution. It outlines how the LGU system has conceptualized its products, clientele, and extension process, and highlights important events that have shaped this system and its potential for contributing to agroecological alternatives. It concludes by describing the dynamic relationship among alternative farming systems, alternative agriculture programs, and alternative agricultural scientific knowledge.

The Land-Grant University System: What Is the Product?

The Morrill Act of 1862, awarding "land grants" to the states for the purpose of funding state colleges, was a remarkable political and cultural innovation to make higher education widely available. These state colleges (and later universities) emphasized the agricultural and mechanical arts, and cultivated scientific sophistication in frontier America. The LGUs have been the foundation of public university education in this country. Subsequent federal legislation expanded the educational mission to include public agricultural science and extension services.[15]

The Hatch Act of 1887 authorized federal funding for agricultural experiment stations, launching modern agricultural scientific research. The initial work done by these pioneering scientists addressed fundamental questions of soil, fertility, plant and animal breeding, pest management, and mechanical innovation. Farmer institutes requested these state experiment station scientists provide them advice. When the Smith-Lever Act of 1914 created a national extension service, researchers were freed to pursue their scientific investigations less encumbered by responsibilities to rural communities.

Two examples of plant breeding illustrate the wealth and controversy generated by experiment station scientists. Jack Kloppenburg relates the history of scientific plant breeding in the United States, and the critical development of hybrid corn.[16] This innovation led to massive increases in corn yields, fostered the agricultural productivity that led to rural migration to urban America, and sparked the adoption/diffusion research agenda to explain how innovations spread. The superior hybrid seeds are sterile, however, and their success transformed seed saving from an on-farm practice to a scientific commercial activity. The success of hybrid corn garnered public accolades and continued funding for LGU research, and laid the foundation for the biotechnology revolution.

California leads the nation in processing tomato production, and during the first half of the twentieth century relied on imported cheap farm labor to harvest this crop. In 1964 Congress ended the *Bracero* program, which had facilitated the importation of Mexican farmworkers. Many California growers were alarmed about the potential for labor unrest so they collaborated with LGU scientists to develop a mechanical tomato harvester. Designing the machine was a relatively straightforward engineering project, but the real breakthrough was the development of a processing tomato that would not bruise. G. C. "Jack" Hanna, a professor of vegetable crops at UC Davis, bred a tomato plant whose fruit would ripen at roughly the same time and was tough enough to be machine harvested: the VF-145. The harvester was manufactured by Ernest Blackwelder, a friend of Hanna. By 1972, instead of 50,000 field harvesters, tomato growers needed only 18,000 sorters to ride on the 1,152 machines. In 1962, roughly 4,000 farmers produced California tomatoes, but within a decade only 600 of these growers were still in business.[17]

Jim Hightower's book *Hard Tomatoes, Hard Times* brought the social impacts of LGU publicly funded research to the attention of a popular audience.[18] This book brought an end to the unquestioned public support for LGUs and inspired a populist agenda that demanded alternative products from them. "From the early 1970s until roughly 1990," Fredrick Buttel notes, "Hightower-style criticism of and activism toward the public agricultural research system focused on a set of closely interrelated themes: the tendencies for the publicly supported research enterprise to be an unwarranted taxpayer subsidy of agribusiness, for agricultural research and extension to favor large farmers and be disadvantageous for family farmers, for public research to stress mechanization while ignoring the concerns and interests of farm workers, and for the research and extension establishment to ignore rural poverty and other rural social problems."[19] According to academic critics, the LGUs were guided by a "productionist" ideology: a deep-seated institutional conviction that increased agricultural production is always socially desirable, and that all elements of society benefit from increased output. This ideology trumpets the public benefits of new technology while obscuring the unequal social and harmful environmental impacts of these technologies.[20]

Yet the LGU system is not homogeneous, and there have always been spaces within it that have attended to alternative approaches to social and environmental issues in agriculture. They have generally been in the minority, because they do not further the productionist ideology. The fruit of their labor often takes the form not of technology but rather of knowledge, especially ecologically informed knowledge. For example, California's Mediterranean climate is conducive to the growing of fruits, vegetables, and nuts. The diversity of crops attracts a diversity of insect pests, and in the nineteenth century UC researchers led the way in the research and formulation of "economic poisons," as pesticides were then called. Charles Woodworth was the first UC entomologist, hired in 1891, and he introduced arsenate of lead, which became the pesticide of choice in California until DDT.

The alternative, minority tradition of California entomology developed in the context of great faith in chemistry. In 1888, a field biologist sent to Australia discovered a naturally occurring predator of the cottony cushiony scale, a pest that was devastating California's citrus crop. Upon importation, the vedalia beetle fully controlled the scale, creating a dramatic public demonstration that the manipulation of predator-prey relationships could provide significant economic benefits to growers. This success gave rise to biological control as a sub-field of entomology. Biological control would provide the foundation for IPM.[21]

The "bug vs. bug" stories sparked the imagination of some growers and their organizations, who demanded pest-control solutions based on biological control.[22] Harry Scott Smith devoted his five-decade career to the study and extension of biological control. In 1923, Smith established the Division of Beneficial Insect Investigations at the University of California's citrus experiment station in Riverside. Smith's division did not merge with the existing entomology department for 50 years, but rather maintained a distinct niche for itself within the LGU, as a division and later as its own department.[23] It became an important seedbed of ecologically informed pest management, with influence felt around the country.

DDT's "miraculous" solution to insect problems pushed ecologically informed pest control into the shadows of the agricultural sciences. Far from being the victims of pesticide technologies, LGU entomologists were for the most part active participants in their widespread deployment, and many of them gained professional stature as a result.[24] The advent of synthetic pesticides led to the capture of entomology as a discipline by scientists who understood their collective mission to be developing chemical technologies to kill pests. Entomologists devoted to their discipline's earlier partnership with ecology remained a small subgroup, oriented toward biological control and its ecological paradigm.[25]

For all their immediate lethal power, chemical pesticide technologies introduced into ecological systems usually create both economic and environmental backlash. While Rachel Carson was writing, a band of entrepreneurial entomologists in California—trained by Smith's Division of Biological Control—developed an ecologically informed approach to manage pests they called Integrated Control, now known as Integrated Pest Management (IPM).[26] Originally developed with insect pests, it has subsequently been expanded to include all pests, including weeds, vertebrates, and pathogens.

The pioneers of IPM used the term "integrated" because they recognized that biological control as a stand alone strategy was often not economically practical, but that when integrated with other ecological tactics, it could be. IPM assumes an understanding of the ecological relationships between crop, pest, natural enemies, human decision making, and the encompassing agroecosystem. It promotes the idea of injury levels, or economic damage thresholds, insisting that not all crop damage is economically significant. IPM allows the use pesticides, but only in ecologically informed ways, guided by data about the spatial and temporal impacts of pests and by an understanding of the ecological consequences of the particular pesticide. Finally, IPM assumes that trained, professional entomologists will work with growers to supervise the dynamics within farming systems, and recommend management strategies based on ecological knowledge.[27]

The IPM pioneers developed these principles from their experience in local collaborative pest-control initiatives, in which growers, professional entomologists (often their graduate students), and research entomologists worked collaboratively to monitor, assess, and manage fields. These cooperative initiatives shaped the social relations necessary to support the implementation of IPM practices.[28] In 1975, two of the pioneers of IPM, Carl Huffaker and Robert van den Bosch, helped draft a plan for UC to create a dedicated IPM program. During the late 1970s, California experienced several pesticide regulation controversies. This created a favorable political climate for this, and the state legislature authorized the UC IPM program in 1979, the first of its kind at a land-grant university.[29] Many farming states have subsequently launched their own programs.

Land-Grant Universities: Who Are the Clients?

Agricultural scientists have to negotiate the "endemic ambiguity" of their profession since the creation of the LGU system.[30] As some of the first publicly funded scientists, they had to establish networks of support for their work. They had to cultivate support from growers and agricultural interests by conducting "useful" scientific work that would result in continued financial sponsorship by elected state officials. At the same time, they nurtured their primary professional allegiance to their scientific disciplinary community, and many early scientists resented the applied work expected of them. Agricultural scientists were able to manage these tensions fairly well until the 1970s, when public critics began to complain about the consequences of the agricultural system that they had designed.

Charles Woodworth recognized that University of California scientists alone could not effectively serve the burgeoning agricultural industry, so he devoted considerable time to working directly with pesticide manufacturers and salesmen. In 1912 he wrote: "I am not sure we all appreciate the tremendous influence the manufacturers and dealers of insecticides are exerting. They are in touch with a hundred growers where an Experiment Station Entomologist reaches one. They have the last word when they furnish the goods just as they are about to be applied. Their advice will go far to confirm or to counteract our recommendations."[31]

Woodworth was instrumental in expanding the clientele of the university of California from the growers and rural communities to include private industries serving agriculture.[32] UC would develop new technologies, which would then be transferred to the private sector, to be used under the expert eye of UC scientists. Thus, by the early twentieth century many scientists had begun to understand their service to agricultural input suppliers having equal importance as serving growers themselves. This kind of collaboration between LGU scientists and agricultural input industries found justification in the productionist ideology, and was very effectively sustained until the publication of *Hard Tomatoes, Hard Times* drew public attention to the way this partnership enriched the wealthy few at the expense of the many poor farmers.

Hightower's work called forth small farmer and rural community activists, who worked together in many states to mount activist efforts to demand major changes at the LGUs. The California Agrarian Action Project (a forerunner of the Community Alliance with Family Farmers) and the Wisconsin Rural Development Center were just two groups giving voice to these concerns. These and other organizations carried Hightower's critique to state LGU administrators and elected officials. These groups were joined quickly by environmental critics of industrial agriculture, particularly in California due to its high pesticide use. In 1978, Robert van den Bosch wrote *The Pesticide Conspiracy*. He described a "pesticide mafia" of agrochemical manufacturing companies, UC and public officials, and large growers who found it to their personal financial interest to promote pesticides, in violation of ecological common sense. In his analysis, growers, farmworkers, the public, and the natural world were the chief victims of this conspiracy. He charged the entire pesticide research, manufacture and use system of suffering from a conflict of interest, and in the tradition of Carson, risked his reputation by going public with knowledge about the abuse of pesticides. His expertise as a biocontrol entomologist added substantial credibility to his arguments.[33]

In 1979, the California Agrarian Action Project (CAAP) joined with the California Rural Legal Advocacy to sue the UC Regents, university administrators, and 300 unnamed faculty and employees for failing to abide by the public good character of the Hatch Act, which authorized the agricultural experiment station. The plaintiffs alleged conflict of interest in the financial arrangements to develop and manufacture the tomato harvester, but made broader claims about the adverse social impacts of mechanization research: frustrating the efforts of farmworkers to organize unions, disadvantaging smaller farmers, and compromising the quality of rural life. When the dust from this legal drama finally settled, UC had won its court case but lost some of its public reputation. Public interest groups were further emboldened to publicize the negative impacts of LGU activities.[34]

LGU administrators generally brushed off these critics, at least initially. During the 1980s, in at least some states, groups inspired by Hightower were able to assemble a sufficient breadth of argument to public officials for sustainable agriculture initiatives. Concomitantly, internal to the LGU system, criticism of research activities emerged. Relative to other research institutions, LGU science was judged too production oriented, too applied, and not sufficiently focused on original discovery.[35] State public funding for agricultural research declined simultaneously with the escalation of federal agency (external) funding, leading to more emphasis on discipline-specific basic research at the expense of applied research to produce knowledge useful for growers.

External and internal forces brought an end to what Fred Buttel called the "Golden Age" of the LGU system in the 1980s.[36] In the 1980s, LGU system administrators, inspired by the success of the National Institutes of Health, identified a new constituency: the life science/biotechnology industries. Figures 2.1 and 2.2 illustrate the Golden Age and the new model. Note the absence of extension activities and farmers in the new

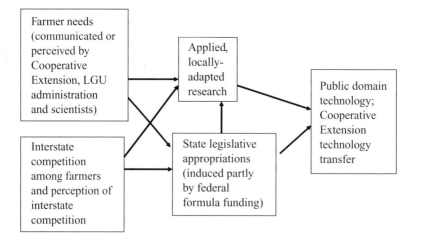

Figure 2.1
Schematic model of the "Golden Age" of LGU research (1940s–1970s). From Buttel 2001.

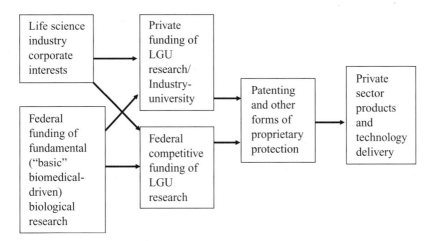

Figure 2.2
Schematic of the "new model" of LGU research. From Buttel 2001.

model. New inventions are no longer to be transferred to agriculture; they are patented and sold to private industry.[37]

The Land-Grant University: What Knowledge Is Extended, and How?

Extension practice is inherently contradictory. It is a pre-meditated, deliberate intervention to achieve the intervener's goals, but it can achieve effectiveness only by inducing voluntary change. Authentic agricultural extension is very difficult to conduct. Extension agents can really only appeal to knowledge to promote change.[38] The design of an extension education program reveals who extensionists perceive their clients to be; it also reveals their assumptions about how their clients learn.

Disagreement about the clientele and organization of extension education work pre-date the passage of the federal legislation creating the extension service.[39] Before 1914, agricultural extension meant either travels by LGU scientists to farming communities (the "demonstration method") or farmers' institutes. In the latter case, rural communities and farmers requested the service of scientists on their terms. The "demonstration method" was pioneered by Seaman Knapp in the American South. His project was to create a corps of scientifically trained experts, based at universities, that would persuade "reluctant" farmers to "modernize" their practices. Arrogance and paternalism tainted this approach.

Farmers' institutes in New York, Wisconsin, and California inspired an alternative strategy of extension grounded in a vision of rural civic life.[40] For example, the "Wisconsin Idea" held that the boundaries of the university extended to the boundaries of the state.[41]

The Commission on Country Life report of 1909 projected this more progressive vision of society.[42] Its author, the plant pathologist Liberty Hyde Bailey, was the dean of the College of Agriculture at Cornell. Influential during the Progressive Era, Bailey combined a faith in scientific expertise and civic ideals. This report included a call for a national cooperative extension system, but it warned against over-reliance on expert knowledge. "Care must be taken in all the reconstructive work to see that local initiative is relied upon to the fullest extent, and that federal and even state agencies do not perform what might be done by the people in the communities. . . . Every effort must be made to develop native resources, not only of material things, but of people."[43] This report proposed a participatory form of extension that would be integrated with moral and political life. Before the passage of the Smith-Lever Act, the Commission on Country Life report contributed to the heated public debate over the direction and the orientation of the proposed national extension system. But the Commission's broader, democratic view lost out to Knapp's approach.[44]

The Smith-Lever Act greatly expanded the LGU project by favoring the approach of Knapp, but extension still struggles with tensions. The discourse of Jeffersonian democracy, founded on the virtue of smallholder farmers, perdured as a social symbol, even as social, economic, and political dynamics shifted power from farmers on the farm to agricultural manufacturing and processing industries, and indirectly, to science institutions. The agrarian populist movements of the early twentieth century tried to stem this loss of power, but were unable to reverse it. Extension practice contains what Scott Peters described as a "dilemma between a scientific impulse to generate and apply expert knowledge and technologies, and a democratic impulse to develop the civic capacities of citizens and communities." Peters continues: "This dilemma has surfaced most strikingly over the years in cooperative extension, as it has tried to balance a 'technology transfer' function with a broader educational mission of helping people help themselves."[45] George McDowell, a specialist in the Cooperative Extension Service, describes this as the "identify crisis"

of the extension. He argues that the public outreach function of the public university, epitomized by extension, has failed to adapt to the new
role of agriculture in society and failed to remain engaged with the issues
facing society at large.[46] He observes that, nationwide, extension has
failed to serve agriculture because it has essentially reduced its role as
serving agriculture, not the public good. He notes a national pattern of
agricultural interest groups "taking extension hostage" by capturing the
resources and work of extension agencies. This is a strategy that has
served them well in the short term, but that ultimately results in a collective loss for the agricultural sector when the public perceives that it is
providing funds for extension with no defensible public purpose. Why
should the taxpayer fund technology transfer to private individuals?
Extension service budgets in the states of Arizona, Oregon, Washington,
Maryland, Texas, Colorado, Oklahoma, and Alabama were all cut
between 2001 and 2004, and Georgia lost more than 50 percent of its
extension service personnel.[47] The UC Cooperative Extension service suffered serious budget cuts: 25 percent between 2001 and 2003. Recall the
schematic for the "new model" for LGU research in figure 2.2, which
omits any role for extension. If the primary research thrust of the LGU
is to patent biological knowledge for private industry, extension services
become vestigial.

Alternative Projects for Alternative Agricultural Knowledge

Against the backdrop of mounting evidence of the environmental and
social consequences of industrial agriculture, recurring indictments of
the land-grant-university system's complicity in designing it, and vexing
questions from state elected officials, LGU administrators in several
states recognized the value of creating programs for sustainable agriculture during the 1980s. At present there are 11 programs, but they vary
significantly in size and mission (see table 2.1). Each represents the interaction between local constituencies advocating for a program and the
context of broader criticism of LGUs. Non-governmental organizations
such as CAFF, and some independent consultants such as Glades Crop
Care in Florida, have played important roles, but none have been as
influential as the publicly funded programs in LGUs.[48] Many of the states
with these programs had farmer institutes prior to the creation of a

Table 2.1
Alternative and sustainable agriculture programs at land-grant universities.

Parent LGU institution	Sustainable agriculture program	Year established
University of California, Santa Cruz[a]	Center for Agroecology and Sustainable Food Systems (formerly the Agroecology Program)	1980
Cornell University	Farming Alternatives Program	1986
University of California, Davis	UC Sustainable Agriculture Research and Education Program (SAREP)	1986
Iowa State University	Leopold Center	1987
University of Maine	Sustainable Agriculture Program	1988
University of Wisconsin, Madison	Center for Integrated Agricultural Systems (CIAS)	1989
Washington State University	Center for Sustaining Agriculture and Natural Resources	1991
University of Nebraska, Lincoln	Center for Applied Rural Innovation (formerly Center for Sustainable Agricultural Systems)	1991
University of Minnesota	Minnesota Institute for Sustainable Agriculture	1992
University of Vermont	Center for Sustainable Agriculture	1994
Kansas State University	Kansas Center for Sustainable Agriculture and Alternative Crops	2000

a. UC Santa Cruz is not a land-grant university. This program emerged in the 1960s from students' interest in small, organic, direct-marketed farming, and from leadership by Stephen Gliessman.

formalized extension service in 1914. All of these programs are quite small, relative to their parent institutions, and generally marginal to LGUs. Many of these programs target smaller, family farmers, and help them with non-conventional crops or direct marketing, and many have advisory boards consisting of farmers and concerned citizens that set priorities for research and programming. These programs average fewer than six employees and less than $1 million in annual budget. Given the political headaches caused by the clamor for alternatives, these programs are really cheap.[49]

The Leopold Center stands out because it was the first at a large, agriculturally oriented LGU, and its creation attracted national attention. In

1998, Practical Farmers of Iowa and the Leopold Center formalized an agreement to forge an intentional multi-year partnership. The partnership seeks to enhance on-farm research and outreach as a vehicle for more effective agroecological strategies and practices. In Wisconsin, the Center for Integrated Agricultural Systems bridged the needs of local farmer networks and the professional goals of scientists by creating a research and extension system titled "radially organized teams."[50] By 1995, Wisconsin had 30 regional networks, 18 of them devoted to grazing. Most of the non-grazing networks were organized around organic production techniques or direct marketing, and many of them worked closely with CIAS.[51] The Kansas program offers support for local networks, many of them graziers, much as CIAS does. Several, such as the Cornell and Nebraska programs, have remained relatively small, and not designed to develop alternative, field-scale farming systems for conventional agriculture.

The two New England programs are small and designed to small farmers in their states. They emphasize direct marketing, organic production, and artisanal food. The Vermont Center for Sustainable Agriculture (based in the Extension System, not the experiment station) is focused on intentional rotational grazing since dairy products are the primary agricultural commodity for this state. The Maine Sustainable Agriculture Program has sponsored integrated crop/livestock research, and tough state-wide nutrient management regulations are stimulating interest in agroecological approaches.[52]

In California, the CAAP/California Rural Legal Assistance lawsuit created a political opening for other critics of UC to act, and State Senator Nick Petris sponsored legislation to create the Sustainable Agriculture Research and Education Program and the Small Farm Center in 1986. Both of these were created against the will of UC leadership, who perceived them as threats to their institutional autonomy.[53] SAREP has played a critical role in agroecological partnerships despite multiple, on-going efforts by UC leadership to suppress it.

In 1984, the National Research Council appointed a committee to study alternative agriculture, naming John Pesek of Iowa State University its chair. The committee discovered that many farmers had already developed techniques that reduced both input costs and environmental harm. A few had invented new techniques, but most had modified their farming systems in light of agroecological principles. The committee identified the two greatest obstacles as the lack of research and the federal farm subsidies that penalized growers for experimenting with alternatives to monocrop production. *Alternative Agriculture* stimulated considerable controversy, in part because its methodology departed from the typical reductionistic approach and in part because it focused scientific attention on agriculture's threats to its own resource base. Its publication was a turning point in the history of American agricultural science. For the first time, the nation's highest scientific research authority accorded a measure of legitimacy to alternative perspectives within agricultural science. *Alternative Agriculture* was published in the year of the *Exxon Valdez* oil spill and the Alar pesticide scare. Although its criticism of dominant agricultural science was only implicit, many LGU scientists felt attacked yet again, but others responded with cautious curiosity.[54]

The same year *Alternative Agriculture* was published (1989), the NRC commissioned a new committee to study the science, technical tools, and policies needed to protect soil and water quality. Four years later, *Soil and Water Quality: An Agenda for Agriculture* called for a systems approach to prevent pollution while protecting farming productivity.[55] It argued:

> . . . integrated farming system plans should become the basis of federal, state, and local soil and water quality programs. . . . Inherent links exist among soil quality conservation, improvements in input use efficiency, increases in resistance to erosion and runoff, and the wider use of buffer zones. These links become apparent only if investigators take a systems-level approach to analyzing agricultural production systems. The focus of such an analysis is the farming system, which comprises the pattern and sequence of crops in space and time, the management decisions regarding the inputs and production practices uses, the management skills, education and objectives of the producer, the quality of the soil and water, and the nature of the landscape and ecosystem with which agricultural production occurs.[56]

This report became the most influential agricultural environmental resource protection agenda during the 1990s. It did not explicitly define "integrated farming system," admitting that variability in cropping systems, ecosystems, and regional contexts made a singular definition impossible. Also left unaddressed were questions of extension program and practice. It did not point to any working models.

Conclusion: Integrating Alternative Farming Systems into the LGUs

During the 1980s, Jim Cannell, Jim Brown, Dick Thompson, and California's partnership pioneers asked their state extension agents for help with issues they thought merited scientific attention, but they found the university had little to offer them. They did not want more technology transferred. They wanted a different kind of science for an alternative kind of agriculture. They needed a different kind of program for an alternative kind of extension.

The land-grant university is an extraordinary scientific institution, remarkable for its size and impact. Unfortunately, it has tended to conceptualize it clientele rather narrowly, and created knowledge and technologies that have benefited a limited portion of the farming community. With the help of Latour's interpretive tool, we can see that LGU scientists and administrators have not paid sufficient attention to developments outside the "colleagues" and "content" loops (loops 2 and 5). Agricultural science's knowledge about undesirable consequences of industrial practices was being circulated outside these loops, by critics such as Carson, Hightower, groups they inspired, the USEPA, and the National Research Council. They argued for alternatives to the dominant production systems.

Advocates for alternative agricultural science and extension programs have generally had to overcome deep-seated institutional resistance, and when they have succeeded, these programs have remained rather small. These programs understand small and medium farmers as their primary clientele, so their marginal political power should not be surprising. Many suffer serious and sustained budget cuts. Nevertheless, they have facilitated some scientific experts stepping out into the field to help develop agroecological alternatives, and the limited entry of agroecological approaches into the dominant agricultural science institution in the United States.

3

Cultivating the Agroecological Partnership Model

The almond orchard was full of life. Glenn Anderson managed it so that it would be. He believed that increasing the diversity of ecological organisms in his orchard would, over the long run, please his almond trees. He grew beautiful cover crops of clover and vetch, and nurtured a wide range of beneficial insects, including convergent ladybird beetles (*Hippodamia convergens*, a relative of the vedalia beetle), lacewings (*Hemerobius* spp. and *Chrysopa* spp.), assassin bugs (*Zelus* spp.), and many species of spiders. His neighbors thought Anderson was crazy letting all those weeds overrun his orchard, but over time they began to notice the diversity of the insect visitors he was attracting.

The Anderson orchard was the first anchor point in a web, capturing the interest of a wide range of growers, scientists, public officials, and environmentalists. A single orchard would have been an anomaly, but paired with his brother's neighboring conventional orchard and observed over many years, this orchard persuaded many that ecological organisms and relationships could provide economic benefit. The creators of the Biologically Integrated Orchard System (BIOS) spun a network out from Anderson's almond orchard.[1]

Farm Advisor Lonnie Hendricks was intrigued by the multi-year comparison data Anderson and his brother had brought to his office. Hendricks had been interested in biocontrol and in integrated pest management ever since the beginning of his career. He had traveled the San Joaquin Valley with Robert van den Bosch, releasing walnut aphid parasites, and he had shared his perspectives with growers and pest-control advisors at monthly IPM breakfasts in Merced County. By the time the Anderson brothers came to him, in the late 1980s, Hendricks was established in his profession and in his dedication to IPM. The funding he received from SAREP for the Anderson brothers' study paid for his research expenses, and he reported provocative results.

Bob Bugg's interest in the ecological dynamics on organic farms dated back to the late 1970s. An entomologist with an agroecological bent, Bugg had heard of the Anderson brothers' organic/conventional comparison through his work at SAREP. In 1992, Rick Reed of the Community Alliance with Family Farmers came to his office and proposed they put together what he described as a "Rolls Royce extension project" for reducing pollution from almond orchards. As program director for CAFF, Reed needed a project that would serve small growers but would have a broader impact on agriculture. CAFF understood that its claims about the viability of alternative agriculture were unpersuasive without a viable example. After tossing around a few ideas, Reed and Bugg recalled the Hendricks study in Glenn Anderson's orchard. So they paid it a visit.

The Community Alliance with Family Farmers blended political advocacy with outreach to ecology-oriented and family growers, drawing these priorities from its two parent organizations. Students and faculty at UC Davis began the California Agrarian Action Project in the 1970s because they were angry with what they perceived to be indifference to labor, environmental, and consumer issues in the UC agricultural research agenda. CAAP was one of the plaintiffs in the tomato harvester lawsuit. The California Association of Family Farmers was founded in 1983 to address the needs of small growers in the San Joaquin Valley. They worked together on several projects of mutual interest during the 1980s, including advocacy for the creation of SAREP to serve their constituents. In 1993, they joined forces to create the Community Alliance with Family Farmers, keeping the acronym CAFF.[2]

One of the earlier projects these two parent organizations had worked together on was the development of a research agenda to serve the needs of small and environmentally oriented growers. After the National Research Council published *Alternative Agriculture*, two California ballot propositions were prepared to provide funds for alternative agricultural research, and the possibility of change was in the air. The two organizations of CAFF wanted a research agenda ready should funding for alternatives become available. With other NGOs, they consulted forty small growers to create a research agenda titled "Farmers for

Alternative Agricultural Research." This recommended that UC and others do the following:

Pursue new research on ecologically sound farming systems that addresses problems identified by farmers and that emphasizes on-farm, farmer-oriented, and systems research.

Collect, evaluate, and organize information about existing alternative practices, especially practices farmers have developed and are using on their own farms.

Actively disseminate information about alternative practices to farmers. Strengthen communication between the farmer and consumer through a new public outreach program.[3]

Neither ballot measure passed, but the project drew people and attention to the potential of alternatives.

The two organizations continued this conversation through the "Lighthouse Farm Network," monthly breakfast meetings with member growers and agricultural professionals. They adapted this idea from a group of small growers in Fresno County, the California Clean Growers Association, formed in 1988 to help, educate, and support local growers trying to practice what they called "natural farming." They grew peaches and other perennial crops near Fresno.[4]

Reed visited the collaboration between Jeff Dlott, a graduate student conducting applied research with California Clean growers. Reed recognized that their partnership manifested the kind of alternative approach to knowledge generation recommended by "Farmers for Alternative Agriculture." This proto-partnership was in the back of their minds when he and Bugg paid their visit to Anderson's orchard and proposed extending the comparison study to more growers.

In one sense, almond BIOS was merely the next step for Hendricks in his development as a scientist and extensionist, building on years of research into the application of biological control and IPM. He never claimed that a bio-diversification strategy alone could eliminate the need for agrochemicals, but he insisted that growers who plant cover crops, who avoid hard sprays, who think sanitation, and who reduce dust and therefore mites are better farmers and can benefit economically from ecological activity. Hendricks's support of BIOS was critical to persuading local growers that the partnership was legitimate and worthy of their time and investment.

The decision to enroll conventional growers but extend production practices derived from an organic orchard was remarkable. Each successive year of Hendricks's study of the organic/conventional comparison had revealed that Glenn's organic methods were viable. Unlike most Farm Advisors, Hendricks was comfortable with participatory research and extension. He made more modest and nuanced claims relative to other BIOS creators, but felt that partnering to promote BIOS practices made him a more effective extensionist.

UC Farm Advisors dedicate their professional lives to maintaining and enhancing UC's reputation in the farming community. They build on 100 years of socially constructing scientific credibility and legitimacy. Pressure on Farm Advisors from colleagues and UC administrators to defend this reputation is enormous. Losing credibility in the farming community can destroy the career of an individual Farm Advisor, and reflect negatively on the UC extension system as a whole. In situations of uncertainty, Farm Advisors recommend pesticides as a form of risk management, both for the grower's crop and for their own reputations. When in doubt, most growers use pesticides, and Farm Advisors have managed risks to their professional reputation by prescribing them.

Bugg, Reed, and Anderson wanted to do more than demonstrate pesticide reduction. They wanted to spark an alternative vision for agriculture by challenging how people thought about farming. "This is a very charismatic production system," said Bugg in an interview. He continued:

It is a beautiful production system. The medicinal herbs in the understory. It's a really beautiful thing to look at. As Walt Bentley said, 'It's like approaching a trout stream, a premier trout stream.' It's beautiful. And the biology is something that just grabs everybody . . . and then you say 'Wow, a production system that looks like this and feels like this. It vibrates like this, also has these great yields.' You say 'Oh my gosh! What have we been missing?' And that gets people excited.

Almond BIOS provided outreach and support for growers as they experimented with agroecological methods taken from Anderson's organic orchard. The creators of almond BIOS built the BIOS model of extension, understood as both content (table 3.1) and process (table 3.2) and flavored by an agricultural populist discourse. Almond BIOS enrolled 26 growers in the initial Merced County project, and helped them establish side by side comparisons of their standard farming practices with a

Table 3.1
Practices promoted by BIOS. Sources: Broome et al. 1997, Pence 2001.

Component of farming system	Practice promoted
Soil management	Compost and cover crops
	Leaf tissue analysis
	Reduced nitrogen application
	Split nitrogen application (according to crop needs)
Insect pest management	Frequent field monitoring
	Bt sprays
	Beneficial insect releases
	Elimination of dormant OP sprays
	Orchard sanitation
Weed management	Winter cover crops serve as standing mulch to reduce summer weeds
	Glyphosate in tree rows
	Reduced pre-emergent herbicides
	Reduced herbicide use in general
Disease management	Conventional fungicide applications
	Careful irrigation management
	Varietal choice
Vertebrate management	Bat and owl boxes
	Trapping

"BIOS block" of about 20 acres of alternative practices growers selected from a menu. This went much further than promoting a static list of Best Management Practices; it tried to capture the imagination of growers by helping them perceive the beneficial activity of ecological organisms, and by encouraging growers to trust their own observation and experience. It required growers to undertake some risk, but provided support for their active learning.

Almond BIOS facilitated social or group learning about the interactions between components of their farming system. CAFF set out to challenge conventional extension practice by re-imagining how traditional participants related to each other in the generation of agricultural knowledge. The BIOS creators espoused a collaborative extension model that facilitated growers exchanging knowledge derived from their farming experience. This challenged the scientific authority of UC Cooperative Extension. Bugg, Reed, and Anderson wanted to help growers reduce

Table 3.2
BIOS extension processes. Source: Dlott et al. 1996.

Component	Strategy
Research	Growers establish comparison blocks to evaluate suites of practices in their own specific agroecological context
	Research scientists investigate specific agroecological practices with participating growers
Extension	Assembles a consortia of growers, PCAs, UC scientists and Farm Advisors, and government agencies to provide a network of support, technical assistance, and financial incentives
	Emphasizes a participatory learning wherein growers, scientists and PCAs share their experiences and insights
	Uses diverse, experientially-oriented social learning tools, especially field days at BIOS farms
	Provides custom support tailored to individual growers' needs
Technical assistance	Customized farm management plan
	Coordinated program of pest monitoring
	Regular updates about the project through newsletters
	Regular information about the agroecological conditions of the orchard
	Cost-share assistance for resource conservation products through USDA

their reliance on pesticides, but they also wanted to defy the conventional wisdom of UC Farm Advisors, whom they saw as actively promoting their use.

Reed, Bugg, Hendricks, and Anderson formed the first management team to select the growers most appropriate for the first year of almond BIOS. They chose growers sufficiently skilled and dedicated to the project. The management team visited individual farms and helped growers determine what practices to try in their own comparison BIOS/grower standard blocks, and helped growers access the knowledge and resources they needed to succeed. At the recommendation of Hendricks, they reached out to pest-control advisors. A few independent PCAs were attracted to the BIOS approach, and helped recruit other growers. (See figure 3.1.)

Figure 3.1
Independent Pest Control Advisor and BIOS participant Cindy Lashbrook
inspects an almond leaf for pest damage.

To fund almond BIOS, Reed contacted Augie Feder, a new USEPA staff
member in the San Francisco office, fresh out of a master's program
in Environmental Science. The agency hired Feder because its analysis
had identified agriculture as the top source of California's unmitigated
pollution. Regional USEPA leadership recognized that their traditional
existing media (air, water) and category (pesticides, solid waste) programs were not sufficient, so they wanted to try a new approach. Feder
had heard of Hendricks's study and recognized it represented the kind of
systems-based pollution-prevention approach the USEPA needed to
support.

Feder encountered BIOS with knowledge of the latest science policy for
addressing agricultural pollution, but also the power of the problem of
the organophosphate diazinon. Originally registered as a pesticide in
1956, diazinon became one of the leading causes of acute pesticide poisoning for humans and wildlife nationwide in the 1980s. A granule is
enough to kill a small bird. It is highly toxic to beneficial insects and hon-
eybees. It has the second highest number of bird kills reported killed by
pesticides. It was and is a priority pollutant in the western United States.

Almonds are the number-one crop nationwide using diazinon, followed by prunes, even though only a fraction of growers apply it annually. Diazinon use was and is a problem for conventional almond growers and regulators, not to mention birds and aquatic organisms.[5]

In the 1970s, UC had promoted the use of diazinon in IPM. When DDT was banned, many growers had simply replaced it with in-season organophosphate sprays, exposing farm workers and wildlife to its acute toxicity and disrupting the summer activity of beneficial insects. A dormant season application of diazinon reduces some of these risks, but when applied in the wet California winter, it is a vagabond chemical, readily migrating to streams, rivers, and ultimately to the San Francisco Bay/Delta before it degrades. In 1987, scientists first identified diazinon here as a primary water pollutant, and traced it back to orchards and alfalfa fields. Simultaneously, another group of scientists discovered that diazanon and other agrochemicals were transported and concentrated by winter fog in the Central Valley, implicating them as public and environmental contaminants.[6] In 1992, the USEPA cancelled ethyl parathion, the other organophosphate used in dormant sprays, so diazinon use among almond and prune growers actually increased. Pesticide policy critics argued that simply swapping new pesticides for older ones as their environmental problems became known was an endless holding pattern, and failed to address the fundamental problems of how growers and agricultural scientists conceptualized the role of pesticides in farming systems.

Diazinon's environmental behavior focused regulatory agency interest in alternatives. Feder provided some initial funds for almond BIOS, and then when the first year's results indicated dramatic reductions in agrochemical use, he argued successfully that support for this kind of community-based, grassroots project achieved the agency's pollution-prevention goals far more efficiently than regulatory actions. His supervisors could not resist his claim that voluntary initiatives were a more efficient strategy to prevent pollution and achieve agency goals.

Almond BIOS conducted agricultural extension in parallel to—or competition with—conventional UC efforts. Many UC Farm Advisors did not take kindly to a non-governmental organization (NGO) competing with them by creating an "alternative extension model," with its implicit critique. Indeed, several CAFF staffers explicitly attacked UC for failing to support BIOS in particular, and pesticide reduction and small growers generally. CAFF made an early decision to not seek UC's scientific imprimatur on their work. This freed them from having to "prove" BIOS practices worked, but exposed them to charges of promoting bad, dangerous or scientifically invalid information. No Farm Advisor wants to have a reputation compromised by association with "bad science," and many saw BIOS as just that.

At a more fundamental level, the organizational culture of UC agricultural science discounts the social and environmental claims of NGOs on their research and extension practices. Some Farm Advisors saw CAFF as an interloper, stealing away "their" growers, research grant money, and scientific legitimacy. To some, Reed was a carpetbagger and Bugg an ideological scientist. Other Farm Advisors attacked Hendricks's initial study with the Anderson brothers for being "not real research." Farm Advisors pointed to instances in which BIOS promoted environmentally friendly practices that would harm growers' economics. BIOS demonstrated that agroecological partnerships could reduce the environmental impact of agriculture, but not without angering some UC agricultural scientists.[7]

Culturing the Agroecological Partnership Model

BIOS was the first California partnership to explicitly articulate alternative agricultural practices with an alternative extension model. It was not the first partnership, but it demonstrated to many others the possibility of alternatives. The agroecological partnership model became the chief strategy for extending alternative agricultural knowledge in California during the decade following 1993. The balance of this chapter describes the institutions and social circumstances that fostered California's partnerships. It analyzes the environmental regulatory agencies and private foundations that actively supported the agroecological partnership with funding programs and financial resources. I then define the agroecological partnership model in greater detail, and describe its malleability as a mental model. I conclude by analyzing how partnerships have helped reduce the use of pesticides by growers of pears, almonds, and winegrapes.

I define an agroecological partnership to be "an intentional, multi-year relationship between at least growers, a grower's organization, and one or more scientists to extend agroecological knowledge and protect natural resources through field-scale demonstration." Some partnership leaders, including SAREP, describe them as "agricultural partnerships," but "agroecological partnerships" is a more accurate term because they develop agroecological knowledge in the field to protect environmental resources while maintaining grower profitability, all supported by applied scientific research. This model contains traditional elements of extension, but deliberately configures them to more effectively promote agroecological knowledge. As described in the next chapter, the scale of grower, scientist, and organizational participation in agroecological partnerships, plus the degree of entrepreneurial leadership they have invested in them, are without parallel in California over the past two decades. A total of 32 partnerships in sixteen commodities were launched in California between 1991 and 2003 (figure 3.2).

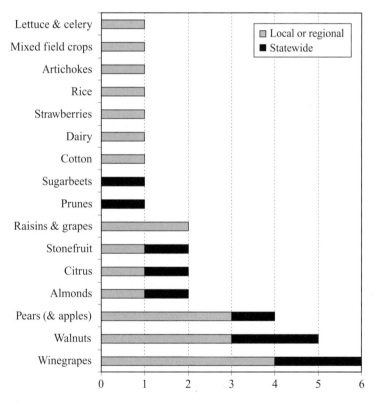

Figure 3.2
California partnerships by commodity and scale.

Expanding BIOS

Even with its epistemological controversies, BIOS demonstrated that alternative extension practices could facilitate alternative agriculture. It opened up the institutional space for other partnerships to emerge through the social agency of growers, agricultural scientists, NGOs, and commodity organizations. These actors brought new programs, funding, and imaginaries to conventional California agriculture. CAFF did not "cause" others to start partnership activities. Other partnerships (the California Clean Growers Association, the Randall Island Project, and the Lodi Woodbridge Winegrape Commission) had been initiated earlier, and partnerships in prunes and in cotton started while BIOS was still in its infancy. The success of BIOS, however, suggested that a different approach to extension would result in socially preferable environmental impacts.

CAFF parlayed BIOS's documented agrochemical reduction to transform it from a pilot alternative extension project into a model. They acquired funding to expand BIOS to other counties, engaged the Almond Board of California with their ideas, advocated for the state to fund BIOS-type programs in other commodities, and encouraged the Department of Pesticide Regulation to undertake its own agroecological partnership programs. CAFF's ambition is manifest in the work plan that was developed during the second year of BIOS:

1. Select one or two commodities with high pesticide use patterns;
2. Identify production practices that could reduce or eliminate targeted pesticides;
3. Create a grassroots outreach program to support farmer experimentation with identified alternative approaches;
4. Work with grassroots groups of growers to identify important areas of research that could reduce pesticide use;
5. Influence the funding decisions of the commodity board.[8]

CAFF relentlessly publicized BIOS, its vision of farming, and its hybrid economic/agroecological advantages, in all manner of farm trade publications. Feder, in turn, helped CAFF obtain a $1.8 million grant from a CalFed, a joint federal/state water management agency. Private philanthropic foundations wanted to invest in BIOS and to be able to claim some success for investing in it.[9] It expanded to San Joaquin, Stanislaus, Madera, and Colusa Counties. It also ran a walnut BIOS partnership in Yolo and Solano Counties.[10]

CAFF reached out to the Almond Board of California (ABC) to gain funding for the *BIOS for Almonds* manual. The ABC provided BIOS funding and a degree of legitimacy with some conventional growers. In turn, the ABC could argue to regulators that it was supporting environmentally responsible research, and to growers it was supporting innovative research. The chair of ABC's production research committee emphasized that the manual represented continued research and learning about almond production, not pesticide reduction, even though the bulk of the manual emphasizes new ways of thinking and farming. In all, almond BIOS operated in five Central Valley counties during the 1990s with $2.8 million in funding. By the time CAFF ended the almond BIOS program, more than 87 growers with roughly 20,000 acres of demonstration orchards—and who managed more than 33,000 acres—had participated in the partnership.[11] Many, many more growers and PCAs had learned about BIOS techniques. CAFF also participated in the almond Pest Management Alliance partnership between 1998 and 2002 (figure 3.3).

CAFF insisted that what they had demonstrated to be successful in almonds could be reproduced in other commodities. The example of BIOS and the advocacy of CAFF stimulated legislators, public agency officials, and philanthropic foundations to fund and create funding programs for more partnerships. The two primary funding programs have been SAREP's Biologically Integrated Farming Systems and the Department of Pesticide Regulation's Pest Management Alliance (PMA) program, but other funders emerged. Informal groups of local growers also adopted partnership strategies.

Creating the Biologically Integrated Farming Systems Program

To legislators, CAFF argued that the BIOS model yielded benefits for growers and protected environmental resources. Reed helped write and lobbied for Assembly Bill 3383, signed by Governor Pete Wilson in the fall of 1994. This subsequently became known as the Biologically Integrated Farming Systems (BIFS) bill, establishing a BIOS-like competitive grants program at SAREP.[12] Bugg and Reed identified unspent funds in the Food Safety account at the Department of Pesticide Regulation, the legislature appropriated $250,000 from it, and USEPA Region IX supplied $420,000 to launch BIFS.

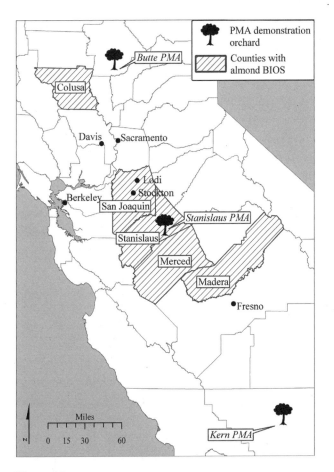

Figure 3.3
Almond partnerships.

SAREP was the logical institutional location for BIFS within UC because the same small growers—and organizations supporting them— had advocated for its creation in the first place. Its mission statement describes SAREP as "a statewide program providing leadership and support for scientific research and education that promotes agricultural and food systems that are economically viable, sustain natural resources and biodiversity, and enhance the quality of life in the state's diverse communities." SAREP's project is to put science into action. While operating firmly in the scientific tradition and methodology, SAREP has an inherently tense relationship with other UC agricultural science institutions because its mission goes beyond basic research and productionism to serve the applied needs of multiple constituencies. BIFS has been one of SAREP's most successful, visible programs, helping it develop a clientele in conventional commodity agriculture. Between 1994 and 2004, SAREP coordinated funding for ten three-year BIFS partnerships, awarding six of them to Farm Advisors, three to commodity and other agricultural organizations, and one to a US Department of Agriculture scientist. BIFS partnerships carried forward BIOS's emphasis on growers' participation and social learning at a local scale, i.e., one county, or two adjacent counties.

SAREP required BIFS grant recipients to create voluntary partnerships that use the BIFS extension model, enrolling local groups of growers in social learning, supported and led by management teams with growers' participation.[13] In one sense, SAREP merely carries forward the recommendation by the IPM pioneers of the 1950s to develop social relations supportive of ecologically informed pest knowledge, monitoring and management. Some BIFS partnerships reproduced the enthusiasm and agroecological vision of BIOS, but others less so.

CAFF pointed to BIOS as a working model of integrated farming systems, and SAREP made similar claims about BIFS being an example of the integrated farming systems approach called for by the National Research Council. Both BIOS and BIFS developed the integrated farming systems model beyond that recommended by the NRC by hybridizing organic pest-management practices with the principles of applied ecology in IPM, which is not surprising given the disproportionate attention to insect pest control in California agriculture.[14] The integrated farming systems approach marks a new stage in the use of ecological ideas in agriculture because its moves beyond IPM's cultural strategies to monitor and analyze the entire farming system so as to manage it in an ecologically optimal way (within the constraints of economic monoculture).

Engaging the Department of Pesticide Regulation

Governor Pete Wilson created the California EPA (CalEPA) during the period when William Reilly and the USEPA were promoting source reduction and pollution prevention. Most of the activities now carried out by the California Department of Pesticide Regulation (DPR) were originally developed by a branch of the California Department of Food and Agriculture and then transferred to CalEPA.[15] One of the DPR's first actions was to create the Pesticide Use Reporting (PUR) system.[16] The DPR conducted extensive analysis of its relationship with its stakeholder groups, which resulted in a pest-management strategy.[17]

The DPR's budget is currently about the same as the amount allotted to the pesticide regulatory programs by the other 49 states combined. Fewer than ten states even have a pesticide regulatory agency outside their departments of agriculture. Many states have no pesticide programs, and devote no state funds to pesticide regulation. The DPR's relative size can be explained by the juxtaposition in this state of the largest agricultural economy with many vocal environmental critics of pesticide use. Many stakeholders make vigorous competing claims on the DPR's attention and resources. The USEPA oversees the DPR's work on pesticide regulation, but also the impact of pesticides on natural resources. Californians for Pesticide Reform and other non-government organizations advocate for more enforcement. Agricultural lobbyists, including the Farm Bureau and the Western Plant Health Association (which represents agrochemical industries), push back and defend the use of pesticides.

The DPR's 1995 pest-management strategy suggested that it could achieve its goals more effectively by developing new educational efforts. Initially, the DPR funded local efforts to demonstrate alternatives and commodity-wide pest-management planning, which would identify existing knowledge not fully used, and gaps in applied research. When the state legislature appropriated its funds generated from pesticide taxes to pay for the BIFS partnerships, the DPR was further motivated to launch its own partnership funding program.

In 1997, Jean-Marie Peltier was recruited from the California Pear Advisory Board, which had been a partner in the Randall Island Project, to become Deputy Director of the Department of Pesticide Regulation. Peltier understood the benefits the pear industry had garnered through this partnership. Her leadership at the DPR emphasized the public/private industry partnership approach. The DPR launched the Pest Management Alliance (PMA) program in 1997 to help industry groups, such as commodity organizations, address pest management on a regional or a statewide scale. That was an ambitious scaling up from previous pest-management planning efforts. The DPR wanted to facilitate the implementation of reduced risk practices by emphasizing the whole pest-management system while building stronger relationships with industry groups.[18]

The Pest Management Alliance program is the first and only state pesticide agency-initiated pest-management extension effort in the United States. Between 1998 and 2002, the DPR sponsored eight PMA grants that qualify as multi-year partnerships.[19] In contrast to the place-based BIFS model, which emphasizes growers' participation, the PMA partnerships are based on the public/private industry partnerships developed between the USEPA and manufacturers using the USEPA's language of risk assessment and reduction. The PMA program was designed to work extensively with commodity groups on a statewide basis to focus applied scientific research on pest management. Before applying for a PMA grant, a commodity group is required to undertake an evaluation of an existing pest-management system. PMA partnerships provide funding and direction to commodity organizations, which recruit UC Farm Advisors, who work with individual growers to manage comparison blocks with "soft" pesticides. Farm Advisors usually pick growers whom they have known for years or decades.

The PMA program does not articulate any alternative strategies for developing or extending alternative knowledge. The PMA partnerships insert newer, alternative, soft pesticides into conventional models of extension, which in some cases have helped reduce pesticide use. The program carries forward the assumption that innovative, alternative practices are developed by UC scientists, not growers, and that growers participate in partnerships through commodity organizations. Some commodity organizations, however, have adopted an integrated farming systems approach.

The DPR and PMA partnerships cope with the same UC bias against practical, applied environmental problem solving research that CAFF and others had noted. UC leadership had complained that CAFF was taking resources for research and extension that should rightly flow to the university, that all SAREP's funds should be under their direct control, and that state dollars used to support PMA research should properly be allocated by UC. DPR staffers note that most of their scientists are the product of the UC system and many have been trained in the

UC tradition of IPM, and that almost all of the PMA grant monies flow back to UC research institutions anyway.

The FQPA as External Stimulus for Partnerships

In 1996, Congress passed the Food Quality Protection Act, the first major revamping of federal pesticide laws since the creation of the USEPA. It established a thorough review of pesticides and threatened to ban organophosphates, causing considerable anxiety among growers and agricultural organizations.[20] The FQPA provided an external stimulus for agriculture to develop alternatives to traditional chemical pesticides, and boosted interest in agroecological partnerships. Perhaps most significantly, it embraced a systematic risk assessment, requiring risk analyses to look at cumulative effects of similar chemical hazards.

The FQPA also created funding for alternative pest-management approaches through the USEPA and the USDA, which have funded various partnership activities across the country seeking alternatives to hazardous pesticides. As figure 3.4 shows, ten of the 32 California partnerships were launched within 2 years of the FQPA's passage, six of them funded by the DPR. The interest in alternative agriculture dating back two decades prior to the passage of the FQPA, supported by the theoretical knowledge of the two National Research Council reports and FQPA funding, now found expression in agroecological partnerships.

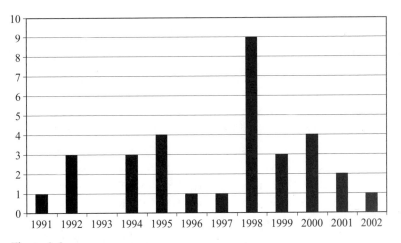

Figure 3.4
California partnerships, by year of initiation.

Table 3.3
Factors stimulating the development of partnerships, as identified by California's partnership leaders. Source: Interviews with agroecological partnership leaders (N = 32). Multiple responses possible.

Farm Advisor, commodity organization, NGO pursues funding	26
Environmental regulatory pressure	22
Group of growers want to pursue agroecological alternatives	15
Development of resistance to pesticides	9
Recognition of public dissatisfaction with existing agricultural practices	9
Anticipation of food-safety concerns	3
Economic opportunity	2

Passage of the FQPA represented a window of opportunity for grass-roots activists, scientists interested in alternative agriculture, entrepreneurial private foundations interested in agriculture, and agency administrators to yoke the environmental and policy problem of organo-phosphate use with the "solution" of agroecological partnerships.[21] California's agroecological partnership leaders reported that regulatory pressures were the second most common motivation for undertaking partnerships (table 3.3), and observers noted similar trends nationwide. The FQPA started many growers' organizations thinking about production practices without organophosphates and other highly hazardous agrochemicals, should the USEPA cancel their registrations. The passage of the FQPA signaled the end of an era. When a pesticide registration had been cancelled, it had been routine for growers to simply substitute another hazardous pesticide, with a slightly different chemical formulation. The threat of OP loss stimulated scientists and growers' organizations to seek out alternative strategies for developing new practices, and new ways of extending those practices.

Building on his experiences with the California Clean Growers Association and other partnerships, Jeff Dlott helped the Pew Charitable Trusts (PCT) develop several agroecological initiatives. PCT had become interested in agroecological partnerships through its financial support of BIOS. Pew hired Dlott to design a partnership for Sun-Maid raisin growers. Based on this success, Pew then asked Dlott to draft a plan for a national agroecological partnership resource center, integrating social, economic, and ecological science support for partnerships. In 1996, PCT established the Center for Agricultural Partnerships (CAP), which has initiated eight partnerships across the United States. Two of them were in California: one with Salinas Valley lettuce growers and another promoting pheromone technologies among Central Valley walnut growers. Partnerships with Washington pear growers and Michigan apple growers extended pheromone technologies (described further in chapter 6).

The Center for Agricultural Partnerships has been an active advocate for the partnership model across the country, and as a private extension service has had served groups of farmers in several states. CAP partnerships with Minnesota and North Carolina field crop growers addressed nutrient and herbicide management. The Neuse River estuary in North Carolina suffered a series of major fish kills in the 1990s, and the state legislature passed regulations to reduce agricultural nutrient runoff in the watershed. In 1998, CAP launched the Neuse Crop Management Project to help farmers "identify and implement economically sound farming practices to sustain productivity while meeting environmental obligations."[22] CAP convened farmers, crop consultants, grower organizations, and North Carolina State University researchers and extensionists, and the Neuse partnership pulled together one of the largest collections of institutional and individual partners (growers). It created more than 105,000 acres of nutrient management plans, a 23 percent basin-wide reduction in nitrogen fertilizer use, and a 40 percent reduction in pre-emergent herbicides. It was also one of the best funded partnerships, pulling together more than $1.25 million in funding from six foundations and agencies.[23] CAP sees its partnerships as providing additional technical support to a group of growers as they attempt to reduce their reliance on pesticides, and stepping into the gap left behind by the LGU extension system's declining support for production agriculture. In recent years, CAP has facilitated agroecological initiatives with growers groups in Michigan, Florida, New England, Oregon, and Wisconsin, leveraging USEPA dollars to fund alternative, semi-privatized extension efforts. In Florida, Glades Crop Care Inc. has developed agroecological initiatives very similar to the California partnership model.[24]

Other private foundations contributed too. The Kellogg Foundation funded an Integrated Farming System initiative from 1993 to 2003.[25] It funded more than 40 local initiatives for more than $30 million, chiefly in states along the East Coast but also in the Great Lakes Region.[26] Most of these attempted to use alternative and direct markets to create more sustainable agriculture, but about a quarter explicitly tried to change agricultural practices.

Nine California partnerships have emerged independent of any major funding programs. These generally are local initiatives manifesting varied grassroots interests of growers and their organizations, and have for the most part aspired only to local impacts. They will be discussed in subsequent chapters.

Creating a Model

The agroecological partnership model is a socially created mental model, oppositional to mainstream agricultural science, guided by the belief that alternative agriculture is possible. Proponents of the model assert that if extension practices incorporate alternative social relations—growers and scientists working together—progress can be made toward achieving positive results in the field. CAFF was the first to describe this model, but the anthropologist Robert Pence analyzed its dynamics in detail.[27] He described the partnership model as follows:

1. a *structure* of local management teams;
2. a *process* of grower outreach; and
3. a *goal* of reducing agrochemical use by adopting integrated farming practices.[28]

BIOS explicitly injected pro-small-farmer values into its literature and its program, but the agroecological partnership model is remarkable for its flexibility. Other partnerships such as the Randall Island Project and the Lodi winegrape partnership drew from their own experience and combined this with selected practices, language, and ideals from BIOS.

Different approaches to the model should not obscure the shared fundamental goal of preventing agricultural pollution through collaborative research and education among growers and scientists. The model was socially constructed, meaning that practices and social relations were negotiated, developed, and promoted by participants to help others imagine an alternative way of farming. It is, therefore, a mental model put into action, informed by the desire to fuse together improved grower economics, enhanced grower satisfaction, and environmental resource protection. Increased environmental regulatory pressures shape the stage in which agroecological partnerships have emerged, but do not by themselves explain their advent. Partnerships have emerged because participants were convinced alternative practices could be viable, and they committed their time, talent, and resources to make them successful.

As with any shared mental model, different actors find different components of it more appealing than others. Growers are attracted to it because they can save money while improving their stewardship, although the relative emphasis on these goals varies widely by grower. Independent PCAs participate in agroecological partnerships because they believe they can learn how to better serve their growers while fulfilling their professional interest in increasing IPM adoption. Affiliated PCAs participate to learn about new products and how to better serve their growers, although they generally resist new practices that threaten to reduce their commissions on pesticides. Recognizing the increasing environmental regulatory pressure growers face, Farm Advisors participate in partnerships to serve growers' needs while safeguarding their economics, even if they do not advance professionally by doing so. The scientists who participate enjoy applied research. Commodity organization leaders have to negotiate the tensions between serving the growers' expressed economic interest while sustaining a positive image of their commodity's growers to regulatory agencies and the public. Regulatory agency staff welcome the concrete demonstration of alternatives to hazardous or toxic practice and the sense of positive engagement with "stakeholders" that partnerships offer. Partnerships succeed at multiple goals in part because they bundle them together to attract and satisfy multiple participants.[29] The next chapter discusses partners' motivations in greater detail.

Assessing Success

The agroecological partnership phenomenon marks a $10 million experiment by public agencies and private foundations to help California agriculture reduce pollution. BIOS demonstrated the viability of alternative practices to many almond growers, and to the Almond Board, but also showed that specialized pollution-prevention initiatives could work.[30] Relative to almond BIOS, all subsequent California partnerships—with the exception of that in winegrapes—have been modest in their

Counting California's Pesticide Use

When Governor Pete Wilson created the California Environmental
Protection Agency and consolidated all regulatory responsibility for pesticides to DPR in 1991, the department created a system to track pesticide
use and estimate exposure risks. This comprehensive pesticide reporting
system was the first of its kind. Limited reporting to county agricultural
commissioners had been required since the state's designation of restricted
use materials in 1949, but starting in 1990 all pesticide applications had
to be reported. Agricultural leaders initially feared that the Pesticide Use
Reporting (PUR) system would be used to identify individual growers with
patterns of high use, but recently they have recognized that it justifies their
claim that growers rarely use the maximum permitted quantity of pesticides, and in some cases, they can present data reporting the decline in the
use of certain pesticides. Environmental critics point out the 100 million-
pound discrepancy between pesticides reported in the PUR and total
pesticides sold in the state. The PUR documented high hazardous pesticide
use during the 1990s, roughly one-third of the total, although reported use
of the most problematic materials has recently declined.

objectives and in the successes claimed. This section describes how
California partnerships have been assessed and some of the difficulties
this entails.

In their need to continually justify their efforts and funding, agencies
and partnership leaders have undertaken numerous efforts to evaluate
their impacts. This struggle to find "results" is exacerbated by the difficulty of evaluating grower learning about agroecological principles and
methods, which cannot be documented in the same simple fashion as the
"adoption" of an add-on technology like a piece of machinery or an
agrochemical.

Documenting reduction in pesticides has been considered the "Holy
Grail" of partnerships, especially by program staff, dating back to the
late 1990s.[31] Many consider it a necessary "proof" of a partnership's
impact to justify additional support. For some, this is the only objective
means to evaluate partnership impact. SAREP, DPR, and CAP staffers
have devoted considerable efforts to measuring project impacts, chiefly
in terms of agrochemical reduction, but also using other measurements.
Many agricultural organizations actively opposed the creation of the
PUR database, but are now finding it useful to demonstrate that their

growers are using fewer pesticides, or at least fewer organophosphates. The PUR is an enormous database with objective numbers; however, it has major limitations, and it should not be used as the only criterion for evaluating agroecological partnerships or any other effort to intervene in agricultural practices. Weather is the most important factor shaping pest populations, and the PUR cannot distinguish between extension efforts and broader trends in agriculture.[32]

Partnerships Extend Agroecology in Stages

Agroecological partnerships provided material support to enrolled growers, generally for 3–5 years while the partnerships were funded. During that time they provided enhanced pest and nutrient monitoring, new products (pheromones, water monitoring devices, cover crops, beneficial insects), and a social learning process to support the deployment of these. Essentially all partnerships reported pesticide or agrochemical reductions on demonstration field blocks or orchards during the life of the project.[33]

Leading institutional actors in the partnership phenomenon recognized that documenting lasting changes in pesticide use after the completion of a three year partnership would be the ultimate justification for their continued financial support by the state legislature or other funders. Demonstrating that agroecological knowledge and resources provided by a partnership during its funding period is not considered sufficient. Unfortunately, some participant growers revert to the risk management logic associated with pesticides. After some partnerships, the economics of the alternative practices are simply prohibitive, either due to labor or product costs. Simply put, pesticides are cheap relative to the skilled labor needed to gather useful data on the agroecological condition of a farming system. Ecological organisms and relationships in a monocultural farming system do not behave as consistently as do agrochemicals.

"Transfer of Technology" is the dominant pedagogical paradigm for UC Farm Advisors. This approach and the "adoption/diffusion" model are based on assumptions not consistently valid for agroecological partnership activities. In contrast to this, most partnerships focus on learning how to *remove* or *replace* a (hazardous or disruptive chemical) technology instead of *adding* one. Partnerships devote most of their efforts to facilitating social learning about agroecological strategies and practices, not simply new technologies. Partnerships that promote growers undertaking an inventory of analysis of their entire farming system appear to be more successful over time than those that simply propose an input substitution. Partnerships' focus on knowledge and learning distinguishes them from traditional "transfer of technology" extension activities, and makes different demands on extension actors and grower learners. The economics of new methods are still critical, but not absolutely determinative.

Agroecological partnerships re-shape a commodity's practices through three general stages. The first stage consists of initial research into why some growers have had success with agroecological methods. Hendricks's study in the late 1980s performed this function. Several partnerships, such as Rice BIFS and Strawberry BIFS, have conducted valuable research toward this objective, but were not able to improve practices beyond grower participants.

The second stage demonstrates the agronomic viability of new methods. Most partnerships have developed effective practices, but they require more (expert monitoring) labor, or are more expensive. For example, the same pheromone mating disruption practices used by pear growers have not been widely adopted by walnut growers because their orchards occupy roughly three times the volume of pears, and the cost of using pheromones is not economically viable yet. Growers participating in partnerships at this stage are motivated to learn about new methods when they are subsidized or supported, but few are willing to spend much more for them once the partnership has ended. The 2003 BIFS progress report documents the pesticide reductions achieved by its partner growers, and describes the potential commodity-wide reduction were partnership practices to become the norm.

The third stage depends on widespread circulation of knowledge about practices among growers of a specific commodity, but depends on those methods being economically viable. Agroecological partnership activities are not the only reason for pesticide reductions in these commodities. Pesticide resistance, weather, and the economics of new "softer" pesticides are also critically important factors, but partnerships have helped all of these commodities and provided the social relations necessary to support widespread learning about alternative practices and how to use them successfully.

Pesticide Reductions Associated with Pear, Almond, and Winegrape Partnerships

Four commodities evince commodity-wide progress in pesticide reduction, as documented by the PUR, attributable in part to agroecological partnership activities: pears, almonds, winegrapes, and stone fruit. I will discuss the first three because they form the primary case studies for my study.[34] The first two show remarkable declines in organophosphate use. Data also suggest that winegrape growers are using agroecological techniques in several components of their farming systems as a strategy for reducing several FQPA priority materials.

The California almond industry has documented the greatest volume reduction of organophosphate use, as illustrated in figure 3.5. Some of this reduction is attributed to growers switching to pyrethroids, which are less hazardous to mammals and somewhat less disruptive of beneficial insects, but are acutely toxic to aquatic organisms. Thus, even though the almond industry has reduced its use of organophosphates and their associated environmental hazards, some new environmental risks are associated with pyrethroids. The size of this voluntary reduction has generated more research into the almond industry's PUR records than

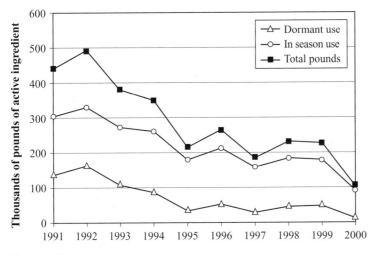

Figure 3.5
Pounds of organophosphates used on California almonds, 1991–2000. Source of data: Elliott et al. 2004.

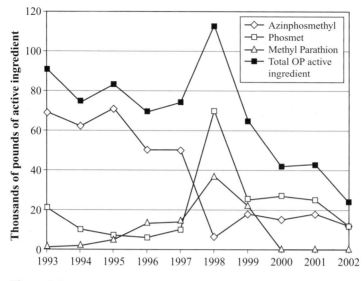

Figure 3.6
Pounds of organophosphates used on California pears, 1993–2002. Source of data: Pesticide Use Reporting database. Data extracted, analyzed, and provided by Bob McClain of the California Pear Advisory Board. The volume of pheromones is measured in fractions of ounces, not pounds, and does not register on this chart.

any other commodity.[35] These researchers explain this reduction by pointing to the economic advantage of new pyrethroids, fluctuations in weather, as well as partnership activities. My focus is to explain the social relations that partnerships fostered as one of the factors contributing to this organophosphate reduction (especially chapter 5).

Pear growers have reduced organophosphate use more rapidly than any other commodity in the history of California agriculture by substituting pheromones, as shown in figure 3.6. Codling moth resistance to organophosphates gave a strong impetus to these efforts, but the economic advantage of pheromone mating disruption became undeniable by the late 1990s, according to Pat Weddle. Figure 3.6 suggests that as pesticide resistance began to appear in the late 1990s, some growers tried switching from azinphosmethyl to phosmet, but then found pheromone mating disruption to be more effective and economical. Methyl parathion was cancelled for use in pears in 2000. Organophosphate use in pears in 2002 was 18 percent of 1998 levels. This is the most rapid reduction in organophosphate use in the history of California agriculture. The initial success of the Randall Island Project was reproduced by three other California pear partnerships, especially the PMA. These provided networks of expert scientific to help growers and PCAs effectively deploy the pheromones necessary to support this organophosphate reduction. Similar successes took place in partnerships in Washington, Oregon, and Michigan tree crops susceptible to codling moth (see chapter 6).

The winegrape industry uses fewer organophosphates relative to almonds and almost all other perennial crops. Winegrape growers are fortunate to have a crop with few cosmetic concerns, generally lower pest pressure, and many alternatives to organophosphates to treat their pests.[36] Winegrape partnerships have focused their attention on the integration of their farming system components, more than other commodities. The winegrape industry is also highly organized into production regions, and partnership activities have taken place in the context of these existing social networks (as discussed more fully in chapters 6 and 7). The winegrape industry does not show statewide declines in organophosphate use, but this does not serve as a valid indicator of pesticide reduction as it does in other commodities. Several researchers have documented the impressive reductions of several (non-organophosphate) FQPA priority agrochemicals in three regions where winegrape partnerships have been active.[37]

In sum, agroecological partnerships have played a substantial role in reducing pesticide use in these commodities. It is impossible, however, for partnerships to claim their activities alone are responsible for these reductions. In general, growers are motivated by cost cutting opportunities (especially if they reduce the need for hazardous agrochemicals), and scientists are interested in practices that reduce the need for hazardous agrochemicals (especially if they are cheap and practical). The simple adoption/diffusion model used to evaluate add-on technologies cannot capture the complexity, multiple criteria, and institutional resistance encountered by agroecological extension initiatives like partnerships.

Conclusion

The morning before I interviewed Glenn Anderson in the spring of 2003, I had spoken with a University of California Farm Advisor. Even though he had been skeptical of the pesticide-reduction goals of the almond PMA partnership, he had agreed to cooperate with the Almond Board's initiative. After several years participating in the almond PMA, he concluded that it was in fact possible to dramatically reduce reliance on dormant organophosphates, and in some cases eliminate them. When I related this to Anderson, he gave me a look of bewilderment. He could not understand why this scientist was only now coming to the conclusion that organophosphates were not essential to almond production, 10 years after Hendricks's study and the launch of BIOS. His orchard was 20 years old, had never received an organophosphate treatment, and was profitable. Why did the implications of agroecological knowledge seem so slow to penetrate the UC agricultural science institutions?

BIOS was the best funded and the most provocative agroecological partnership. It wove together all the components that came to define the agroecological partnership model: grower-generated agroecological knowledge, supportive PCAs and Farm Advisors that facilitated social learning, applied scientific research to support alternative practices, supportive work by grower-oriented organizations, and emphasis on less hazardous technologies. Most perennial crop growers in California had heard of "the BIOS model" or "agricultural partnerships" by the end of the decade, especially as their commodity organizations jumped in to partnership activities.

Agroecological partnership leaders developed their model informed by the circulation of ecological knowledge. BIOS creators built the model based on critical, comparative agroecological knowledge emanating from Glenn Anderson's orchard, verified by grower experimentation replicating his success. BIOS was based on selectively emphasizing this knowledge to provide superior grower satisfaction in comparison to the conventional, chemically intensive practices recommended by most UC Farm Advisors. BIOS parlayed initial success into a working model and promoted it broadly, competing with and criticizing conventional UC extension. BIOS set out to prove wrong assumptions about the inevitability of chemically intensive agriculture. California has a greater diversity of agroecological partnerships due to its diversity of crops, and the uncertainties posed by the FQPA. Other states have independently developed initiatives with many of the same partners, funders, characteristics, and goals.

The applied character of agroecological partnerships expands the menu of the scientific activities conducted by its participants. This kind of agricultural science is not restricted to original laboratory discovery for consumption by other scientists. Growers, scientists, advocates, industry leaders, and public officials perceived the problem of agricultural pollution linking field, scientist, growers, market, and public—all five loops of Latour's model. Many of these actors fashioned feedback loops to circulate information about public perception of pesticides in agriculture to scientists and growers, weaving together what others had perceived to be distinct loops of knowledge. Partnership leaders held out an alternative vision of agriculture, supported by science with an agroecological orientation, and set about developing and extending the practices needed to support their agroecological imaginary.

Skeptics have dismissed the importance of the agroecological partnership model, claiming it is nothing more than the incremental progress of agricultural science responding to environmental regulatory pressure. Indeed, the agroecological partnership phenomenon would be largely invisible to the positivist. Partnerships describe the beneficial interaction between ecological organisms, only tacitly referring to ecological principles. Scientists and growers have experimented together with ecologically informed practices for more than 100 years in California. Extensionists have promoted resource protection activities and growers have participated in field-scale demonstrations. These activities are not new, but, as the next chapter indicates, these partnerships succeed because they foster the requisite social relations. It is largely through these developments that the agroecological partnership model reveals the imaginary that underlies how agroecology moves from theory into action.

Agroecology in action is much more powerful and persuasive than ecological knowledge in a laboratory or textbook, confirming Latour's recommendation to look for circulating knowledge. This chapter demonstrates how partnership leaders circulated ecological theory into action out in the field with growers, and then pumped it back to into science and regulatory institutions. The BIOS partnership exemplifies how agroecological partnerships vascularized three of the outlying loops in Latour's model: nature, scientists, and agricultural clients. The next chapter investigates the motivations of these actors and their reasons for enrolling in these kinds of science projects.

4

The Partners

The Gorilla Goes to the Farm

Karl Kupers recognized that his farming system was challenged on both sides: economics and production practices. His family had grown winter wheat on roughly 5,000 acres of leased land for several generations in east central Washington. He had made a satisfactory profit, but depended on federal subsidies for generations to do so, and he knew these weren't going to last forever. An alternative economic strategy was going to become crucial. At the same time, environmental problems associated with conventional wheat production in some parts of this region were becoming undeniable. Spokane environmental groups were up in arms about the air pollution from burning stubble. Wind blowing across exposed soils during the summers caused occasionally hazardous dust storms. Rain on exposed fields resulted in gully erosion, and deposited silt in some streambeds. Nitrate from fertilizers was finding its way to groundwater, and was coming under regulatory scrutiny. The Seattle office of the USEPA had identified the Columbia Plateau of central Washington as home to the region's worst unregulated environmental problems, and the chief source was agriculture. Kupers knew that "business as usual" was not sustainable, and he refused to be passive in face of the coming change. He wanted to be someone who would shape that change, both economic and environmental.[1]

Kupers's home in Harrington, Washington is in the intermediate rainfall zone (12–17 inches per year) of the Columbia Plateau, but most crops are grown here without irrigation. Winter wheat dominates production in the region, but this requires careful soil management to ensure moisture is present in the soil for fall planting. Farmers accomplish this by using summer fallow every second or third year, prior to seeding winter wheat. Summer fallow involves leaving the ground unplanted for a year, with repeated tillage throughout the summer to control weeds. This practice creates a "dust mulch," which is generally effective at conserving enough moisture for germination, but it leaves soils exposed prior to seeding the wheat.

During the mid 1980s, Kupers experimented with no-till strategies. "Conservation tillage" is the broad term encompassing a collection of reduced tillage practices, such as ridge till and no-till. Conservation tillage strategies disturb the soil less and leave more crop residue on the surface than does the conventional moldboard plow. With no-till, a specialized grain drill places seed and fertilizer directly in the ground without preparing a conventional seed bed, making it even less disruptive than ridge till. Crop residue remains on the soil surface to reduce erosion by wind and water. No-till systems tend to increase soil organic matter and encourage biological activity in it, depending on herbicides to control weeds and make sure crop plants are able to out-compete them.[2]

Kupers planted canola using conservation tillage in 1986, but the crop failed to establish, so he re-planted 300 acres of perennial, native grasses. He harvested the seeds and sold them to public agencies for habitat restoration, and discovered the potential of an alternative, niche market. Once he figured out a few details, the native grasses seemed to grow rather easily in this area. Over time he noticed that the soil in this field was softer and more friable. Plus, it maintained moisture and seemed less vulnerable to erosion.

Many of Kupers's neighbors also had experienced crop failures with no-till, giving this strategy a bad name. No-till had developed in the wheat and corn/soy farming regions of the American Midwest. It did not translate easily to the Pacific Northwest, however, because this region does not receive summer rain. In 1995, the Monsanto Corporation paid for Kupers and a dozen other innovative farmers to visit Pierre, South Dakota, and Dwayne Beck's research farm to discuss how conservation tillage might be adapted to the Pacific Northwest.

The Dakota Lakes Research Farm was started by a group of farmers who were interested in enhancing the ability of South Dakota State University do applied research. The group of farmers wanted local control of the research; however, this land-grant university did not have the resources to purchase land, so they obtained land and equipment to do research. Since there was not sufficient funding to operate the station, they designed it to have two purposes: research and production. The farm is funded by profits from farming as well as by the university and other granting agencies. The production enterprise is also used to identify research needs, and to test whether techniques that have looked good in research trials can be scaled up for commercial use. Many of the first no-till or conservation tillage efforts in the Dakotas had conserved soil and water, but these had not translated to enhanced yields or profitability. Some research suggested that no-till with proper cultural practices could successfully overcome problems associated with other forms of conservation tillage, so this is the primary focus at Dakota Lakes Research Farm.[3] Herbicides are powerful tools for weed management, but farmers and scientists learned that conservation tillage is much more than using herbicides. Many farmers were captivated by the technology, but discovered that they would have to re-think their farming systems to take advantage of its potential.

Beck emphasized the role of cultural practices, chiefly crop rotation. By managing different crops in different fields and changing them from year to year, farmers capture several ecological advantages. Many diseases and insect pests build up through several years, and crop rotation breaks these cycles. Single crop management at large scales requires farmers to work virtually non-stop while the conditions are right to plant, cultivate, or harvest a single crop plant. By diversifying a farm's crop mix, farmers are able to distribute the work load throughout the growing season.

Learning how to manage soil moisture through cultural practices is critical to no-till, because crop residue acts like a mulch, covering the soil. Much of Beck's work has investigated how to rotate different crops, and how to manage planting intensities of these crops to create optimal soil conditions. Too much moisture results in nitrogen fertilizer leaching into groundwater, or the development of saline seeps. When soils are saturated with water, nitrogen fertilizer migrates to low spots, but in concentrations too high to support crops. They may range in size from a few square yards to tens of acres, and are too wet to allow passage of farm machinery. Too little moisture or nutrients will result in crop failure. Selecting the right crop mix over the course of a multi-year rotation and assessing the right cropping density can result in improved soil function and can improve overall crop yields.

Karl Kupers returned from the Dakota Lakes crash course in no-till bursting with ideas—and enthusiasm. He realized that by managing his agroecosystem so that it mimicked the mix of plants in his local, native ecosystem, he could create a more sustainable farming system:

> You look at that [mix of native species], and say those are the percentages I want in my system. You still recognize that you're going to have 50 to 60 percent of your ground in grain, because that's what grows naturally out here. You're going to have 25 to 30 percent of your ground in cool-season broadleaves, about 10 percent in warm-season broadleaves, and about 5 percent in warm-season grasses. . . .[4]

This mix reflects the moist winters, relatively mild given its northern latitude. Wheat and barley are cool-season grasses. Legumes, mustards, and brassicas occupy a niche in the system for cool-season broadleaves. Warm-season broadleaves are sunflowers, chickpeas, garbanzos, and buckwheat. Corn, sorghum, and millet are warm-season grasses. Kupers said:

> The key point is that it's a complete change of philosophy. In winter wheat/summer fallow [rotations], crops and variety are the first thing I decide. But in this system, it's the last thing I decide. Rotations are for the disease control, for fertility management. Feeding the soil is first, because I'm going to make sure the soil enhances my productivity. [I feed the soil instead of] the crop—a totally different concept. You use rotations, not chemicals, which reduces your cost. You time your crops for equipment and manpower; therefore reduced costs. Those are the key components to making a sustainable agricultural system work.

Once Kupers recognized that it was possible to learn from native ecosystems, he began to see more clearly how federal crop subsidies had distorted the economics of agriculture. He recognized how much of American agriculture farms federal subsidies rather than crops. The chief advantage for Kupers was that a well-designed rotation using no-till reduces weeds and diseases, and by reducing the expense of controlling them, he hoped to come out ahead.

Conventional industrial agriculture sets out to grow a crop with minimal consideration of local conditions. Fertilizers are often applied at a rate that ensures that the crop will never lack them, but this has resulted in the kind of hypoxia, or marine dead zones, described earlier. Weeds, insects, and diseases are entirely predictable consequences of monocropping, and elementary ecology suggests that one should be surprised when agrochemicals are not needed to control them, rather than when they are. In a conventional farming system, these organisms are attacked with additional chemicals, most of which drift from the targeted pests. Indeed, one study found that less than 2 percent of sprayed insecticide actually contacts insect pests.[5] This amounts to treating a symptom with medicine that creates additional problems, requiring additional medicine. All monocrop agroecosystems are leaky, and heavily tilled fields are particularly vulnerable to nutrients leaking into air and water. Continuous cover and rotational cropping do a better job of retaining moisture and nutrients because they mimic natural ecosystem functions. They require higher rates of herbicides, but these farmers believe conservation tillage advances them substantively toward sustainability.

One of the farmers working with Dwayne Beck described the conceptual shift associated with his approach to farming as "a brain-transplant way of thinking." Weeds are no longer a problem, something one needs to kill. Instead they are a symptom of imbalance in the cropping system. If weeds are finding the conditions to establish (sufficient light, moisture, nutrients), they work to fine-tune the crop rotation and intensity to prevent them in the future. This kind of farming requires farmers to learn how to farm with nature, how to re-think the industrial logic of conventional farming.

Kupers recognized the potential for adapting what Beck and the Dakota Lakes farmers had learned to his local conditions, but he was not free to make this decision by himself. He had to negotiate with the ten people who owned the land, descendants of the man who originally homesteaded this area. He had to convince them that his approach was agronomically and economically sound. He presented a multi-year plan for transitioning the entire 5,000-acre operation to no-till, including an alternative economic strategy, no longer based on federal crop subsidies, with a 10 percent cushion. The land owners accepted this, and he converted his entire operation to no-till.

Kupers knew he could learn a lot more if some neighbors conducted their own on-farm experiments, like the South Dakota network. Kupers brought six farmer friends and Diana Roberts, a Washington State University (WSU) agronomist, back to Dakota Lakes the following year. Roberts was part of the Ag Horizons team at WSU, an interdisciplinary, self-directed team of researchers and extensionists working on the Columbia Plateau. Her job description includes educating farmers about improving the sustainability of their practices (economic, environmental, and social). She and the Ag Horizons team recognized the value of farmer to-farmer learning, and understood their responsibilities to include learning together with them.

At Roberts's suggestion, Kupers and the farmers who went to Dakota Lakes in 1996 formed what they called a "support group." Even though some, including Roberts's supervisor, were surprised to hear this psychological term used by farmers, the group was a success. They met several times each year, and shared information about their on-farm experimentation: rotations, cropping density, managing soil variability, controlling water erosion, and marketing. Several members of the group also drew from their experiences at a "Holistic Resource Management" workshop, in which they learned how to make farming decisions consistent with their personal and economic goals.[6]

The support group named itself Annual Cropping, Intense Rotation, Direct Seed (ACIRDS). "Direct seeding" is now the preferred term in Washington State, because it encompasses the use of both low-disturbance (no-till) and high-disturbance grain drills that seed and fertilize in one pass over the ground. Visiting the Dakota Lakes farm inspired them, but they realized that they would have to adapt Beck's research to their own agroecological conditions. Some of these other farmers had also experienced the no-till failures themselves, so they realized several things: they had a lot to learn, there was no one who could tell them what they needed to know, and that the only way to avoid another economically catastrophic failure was to learn from and with each other in a network. They saw the value of direct seeding, but to make field-scale changes was risky. The transition from intensive tillage to a more agroecological approach required more knowledge, and they became a network for generating and exchanging that knowledge.

ACIRDS had seen the critical value of a whole farming systems research farm at Dakota Lakes, and wanted WSU to bring its expertise to bear on their efforts in the intermediate rainfall zone. Research done in other rainfall zones was of limited practical value to their needs. In 1997, Kupers approached Roberts about using the Wilke Farm, an existing research farm in Davenport. WSU had been conducting some research on it for a decade, but never on a farm scale. Roberts consulted with a team of other extensionists in eastern Washington, and they responded with enthusiasm, as did the Wilke Farm committee. Now they needed to find a source of funding for whole farming systems, which was difficult to do within the existing funding sources at WSU. A team representing a funding agency serendipitously visited the area that very same year.

USEPA staff in the Seattle regional office had been talking about the Columbia Plateau's agro-environmental problems throughout the mid 1990s. When they analyzed Washington State air, water, and land pollution—plus habitat loss—they discovered that agriculture was the least regulated source. They also realized the laws authorizing their regulatory activities were poorly designed to address agricultural pollution—most regulations were designed with manufacturing, not agriculture in mind. The Columbia Plateau aquifer was of particular concern because so many rural communities depended on it for drinking water, and it was vulnerable to overuse and agrochemical pollutants. The agency legally had the authority to review all federal actions impacting groundwater quality, and some agricultural leaders feared the agency could interfere with federal crop subsidies, even though this was never seriously considered by the agency. The USEPA ultimately decided to support local efforts to protect the aquifer, but the multi-year review process amplified fears among local residents that this was only the beginning of the agency's efforts to impose stricter regulations in the region.

During the Clinton Administration, the Seattle office of the USEPA secured funding for several community-based environmental protection initiatives, efforts to work cooperatively with local agencies and citizen groups to address environmental problems that are beyond the scope of typical regulatory devices.[7] This approach identifies geographic regions with environmental problems unaddressed by existing programs, and it provides institutional support, coordination, and some funding, for local efforts. USEPA leadership had recognized the limits to the "command and control" regulatory and enforcement approach, and believed that an approach that emphasized institutional collaboration had a better chance of making progress toward environmental goals.

Persuaded by data, the Seattle USEPA office created the Columbia Plateau Agricultural Initiative (CPAI) in 1997, drawing from existing agency staff. Chris Feise, then an extension specialist from the Pacific Northwest land-grant universities serving as a liaison to the agency, insisted that CPAI staff first conduct a listening tour of the farmers and agricultural institutions in the region. He had extensive experience with sustainable farming systems, but he realized how much USEPA staff had to learn about agriculture if this initiative had any chance to bear fruit. Plans for the one week tour included 30 meetings in 5 counties with more than 90 people. For several staff members, this would be the first time ever seeing the Columbia Plateau or visiting a farming operation.

CPAI ran afoul of controversy before the listening tour even left Seattle, however. A pubic information document announcing the initiative was released without being fully vetted by staff aware of the feelings of suspicion held by some of the farmers in the region. In another context, the announcement would have been cheered by a population eager for resource protection initiatives. On the Columbia Plateau, it triggered highly critical publicity, led by several agricultural commodity groups.

The listening tour was a crash course in the sociology of agriculture. USEPA staff listened to grape growers, potato farmers, wheat farmers, commodity commission representatives, county commissioners, wildlife agents, agrochemical sales representatives, extension agents, university researchers, conservation district employees, and staff from other federal agencies. This was a whole new world for USEPA staff. Many of the institutions these people represented were brand new to them, and they began to appreciate the complexity of factors shaping land management decisions and their environmental impacts, and the roles played by economics and public policy in these decisions. They recognized the importance of understanding the social institutions in agriculture. For the first time, some staff realized that crops like potatoes and dry-farmed wheat differ in significant ways, and would require different approaches to address their environmental impacts.

The listening tour heard an earful about the credibility problems of regulatory agencies. Farmers explained some of their fears: that the agency rendered decisions affecting the agricultural community without their input; that these regulations will drive them out of business and a way of life; and that costly efforts to improve their environmental practices will only be rewarded with more regulations. The growers convinced the listening tour staff that the lack of communication between farmers and agency staff would jeopardize their credibility, as well as stable food production in the region.

The listening tour discovered that tremendous changes were taking place in Washington State agriculture, led by bright, educated, knowledgeable, and innovative people. Some farmers were reducing water consumption and addressing the threat of groundwater contamination, and at the same time trying to enhance their profitability, such as by deploying drip irrigation on crops that had never used them before.

The CPAI team had known they were going to have to create new approaches to working with the rural communities, but it was surprised to discover how difficult it was to persuade their colleagues of the merit of a constructive engagement with agriculture. They discovered how deeply entrenched the regulatory and law enforcement approach were in the culture of their own agency. Back in Seattle, they reported the broader implications for their agency's practice of a community-based approach to environmental protection. They explained that some of the complaints against the agency had some merit, and that the agency would have to develop new ways of fostering trust among the people whose activities it regulated. To succeed, the USEPA was going to have to understand much more about the structure, logic, and economics of agriculture—indeed the culture of agriculture. The agency was used to negotiating with industry lawyers, but CPAI was going to require them to get their boots dirty. To build a partnership with farmers, the USEPA was going to have to earn their trust.

Perhaps the most important lesson was appreciating the dignity of the farmers in the region. USEPA staffers discovered that the stereotypes they held of farmers as ignorant, needlessly destructive, or environmentally indifferent eroded in the face of meeting real farmers in specific circumstances. The listening tour did not encounter any farmers environmentally irresponsible, although there was a wide range in their perception of the seriousness of agriculture's environmental problems. The team discovered that farmers for the most part were conscientious land stewards, but that their choices were circumscribed by economic and structural issues in agriculture; some wanted to improve their farming systems, but lacked the power to change industrial agriculture's economic incentives.

One particularly poignant meeting during the listening tour took place in Lincoln County. After a potluck dinner marked by wariness, Karl

Figure 4.1
Growers, scientists, and USEPA staff visit the Wilke Farm for a field day. Photograph courtesy of Diana Roberts.

Kupers came right out and said he did not trust the USEPA. Another farmer said that those who would subjugate first regulate. The head of the listening tour was equally frank in his reply to the farmers, but the honesty helped opened the door for a more substantive conversation. The farmers challenged the USEPA to back up its stated goals by funding progressive agricultural initiatives, and the ACIRDS farmers just happened to have a research plan for the Wilke farm that was the most holistic approach to agro-environmental problems the staffers had seen (figure 4.1). The CPAI team returned to the Seattle office with a much greater understanding of the problems and possibilities of partnering with farmers and their organizations to produce positive change in agriculture. Over the next several years, the Seattle office contributed more than \$600,000 to CPAI-related projects, including the Wilke Farm/ACIRDS farmer partnership. CPAI helped the USEPA learn that progress toward pollution-prevention goals in agriculture was possible.

When I visited Kupers in the summer of 2000, he was waiting for a mechanic to fix a piece of machinery. He took me on a walking tour of a few fields, and showed me what he had learned to see. He radiated enthusiasm for his work with the land. This new way of farming gave him a great deal of satisfaction, partly because it allowed him to farm in a way consistent with his values, and in part because he had largely devised it. He had discovered a way to put his land ethic into action. We spoke of the challenges of creating markets for alternative crops, those that he managed so his system more closely mimicked the native ecosystem.

For ACIRDS farmers, practicing a high degree of stewardship had become part of their culture. These farmers have continued to struggle with very low wheat prices and have been stymied by a lack of rotation crops that are well adapted to this production region. Also, the entire infrastructure for grain crops in this region is organized around wheat production, not a diversity of crops. These producers face a genuine challenge in figuring out how to market these alternative crops so they can receive compensation for the additional effort they expend in their land stewardship. The Ag Horizons team offered scientific expertise, and the USEPA tried to be supportive, but without some major changes in agricultural economic policy, it is not clear how much of an impact ACIRDS's direct seeding efforts will have on regional practices.

Latour's circulatory system of science model helps conceptually organize the actions of these individuals and institutions, and explains why they engaged in agroecological partnership activities (figure 1.3). To implement direct seeding strategies in Washington State required investigating crops, soils, machinery and chemicals, and the behavior of nutrients and water in a specific agricultural region; the participants had to "mobilize the world" to make progress toward a kind of agriculture more consistent with their values. To do so, producers had to tap into the scientific expertise of the Ag Horizons team and Dakota Lakes Farm researchers; these scientists were open to collaboration with farmers and each other in social learning to achieve practical outcomes. To create a more coherent clientele for this kind of knowledge, Kupers and like-minded farmers organized a support group and cultivated interest with the Wilke Farm; ACIRDS became a voice for them and helped them to negotiate with scientists and public agency officials. To fulfill its resource conservation goals and represent the public, the USEPA had to send its staff over the Cascades to investigate the institutions structuring conventional agriculture; the CPAI initiative served to mediate between the public's desire for agricultural goods with environmental protection (another form of mobilizing the world). Agricultural science is more than publishing the results of experiments in scientific journals. It also consists of mobilizing the behavior of many people, technologies, and organisms toward a common goal. This chapter investigates the motivations of the human participants in these kinds of networks.

Why Partners Build Social Networks

Farmers and agricultural scientists, with the help of farmers' organizations and public agencies, have created agroecological partnerships. This chapter analyzes the participation of these four kinds of partners in agroecological partnerships. The most successful partnerships have been driven in part by the initiative of farmers. This chapter analyzes "leading growers" like Karl Kupers, explains how they impact other farmers, and describes a three-fold typology of farm-management styles of participating farmers. It then turns to analyzing the roles and motivations of participating scientists. Agricultural scientists are arrayed in a hierarchy, with LGU research scientists occupying the top echelons, extensionists like Diana Roberts in the middle, and applied, field scientists at the base of the pyramid. Partnerships have had to secure the active contributions of scientists with research and applied skills. Agroecological partnerships mark the entry of agricultural organizations into extension activities, and this chapter concludes with an analysis of their contributions to partnerships. Public agencies concerned with resource conservation have played a role in most partnerships, linking agriculture with public concern about its practices.

This chapter and the next two draw heavily from original research into the organization and practices of California's 32 partnerships, but the findings have broad implications for organizing any agroecological initiative. This chapter explains the activities of participants in partnerships like ACIRDS/CPAI using data from the California partnerships to illustrate the motivations for collaboration. Farmers, scientists, farmers organizations, and public resource agencies exist in every agricultural state, and have conducted agroecological initiatives, some of which rise to the level of partnerships; all can benefit from using a network approach.

Farmers like Karl Kupers participate in agroecological partnerships because they believe they will help them become better farmers, perhaps learn how to save some money, and better manage environmental regulatory and public perception problems. Through their participation, farmers confer critical legitimacy to partnerships and their activities, processes, and goals. Growers have expressed their leadership by identifying problems, proposing collaborative solutions, initiating partnerships, directing existing organizations to address agro-environmental issues, providing a template of new practices, recruiting growers, and advising partnership management teams. The importance of growers' participation in establishing, orienting, and legitimating partnership activities cannot be overstated.

Not all grower participants are equal, however. Most partnership leaders ascribe great importance to the leadership by select growers who have created a suite of alternative practices and offered guidance about these to their peers, such as Karl Kupers. These "leading growers" identified a cluster of motivations for devoting time to partnerships that blended altruism and self-interest. Many partnership coordinators describe this as having leading growers "bring other growers along." Area-wide IPM can benefit everyone, but the initiating grower stands to benefit as well, as did Doug Hemly in the Randall Island Project. An organized group of growers is much more likely to obtain help from researchers and extensionists than disaggregated individuals; partnerships help growers represent themselves, either to the public or public agencies, as Randy Lange and John Ledbetter of the Lodi winegrape partnership discovered. Some growers, especially in CAFF-sponsored partnerships, wanted to farm in accord with agroecological principles, and wanted to help others do the same. In all these cases, leading growers can benefit. Ultimately, Karl Kupers sought out the experience of other like minded growers so that he himself could learn.

Many practices promoted by partnerships were first developed by leading growers who simply decided to reduce their agrochemical use in ecologically informed ways, developed an alternative technique (often in disregard of official scientific recommendations), and made changes in their farming system after several years of experimentation. All the growers described in the narratives opening each chapter of this book undertook on-farm research because they were convinced another way of farming was possible. Partnerships provide social validation and additional knowledge support for these growers. They are small in number, but many partnerships take full advantage of the on-farm discoveries of leading growers. Their innovations earn them the term "template grower," meaning that partnership leaders propose their approach to farming as a model, even if the practices cannot be easily replicated. Template growers provoke a re-thinking of assumptions about the absolute need for agrochemicals among other growers, scientists, and extensionists. These kinds of growers have a special credibility among their peers, even among those who have great respect for the expert knowledge of scientists. This credibility does not always translate into mimicry, however, because other growers and scientists may bracket their novel farming systems as exceptional due to special agroecological traits on that farm, or exceptional grower skills. Even bracketed template growers provoke questions about recommended practices, and demonstrate the potential of alternatives.

The distinction between leading growers and enrolled growers is critical to understanding growers' participation. Enrolled growers are those who have been invited to participate, and generally learn from others, but also contribute valuable practical experience. These growers may have experimented informally with agroecological practices on their own, but they have not engaged other growers or organizations on this subject. Enrolled growers are expected to dedicate a field or block for experimenting with a new practice, learn more about pest or fertility monitoring, and share this information with other growers and extensionists in a structured way. In a majority of partnerships, enrolled growers were expected to experiment with alternative soil and water management techniques, and in a few they conducted some kind of analysis of their farming systems. At public field days, other growers learn about these practices, but those formally enrolled receive special attention and resources from the extensionist. In some cases, they are assisted with cost-sharing for inputs, such as cover crop seeds or pheromones.

Partnership leaders emphasize to growers that participation in partnership activities is voluntary, and that they can withdraw at any time. In California, they express this to growers by telling them they can "spray out" their orchard or field if they so decide that it is economically necessary. Growers' participation among the 32 partnerships varies widely, and takes different forms according to the biological factors of crop production, and the social relations in that commodity and specific partnership.

Twenty-nine of the California partnerships reported formally or semi-formally enrolling growers. They enrolled 24,000 acres in partnerships, and they collectively farmed nearly 250,000 acres.[8] Who are these growers and why do they enter into partnerships? Before turning to different characteristics of growers, let us review some of the general traits of California growers, especially perennial crop growers. California growers pride themselves on being scientific businessmen, but some smaller growers without the access to knowledge, technology and economic resources have contributed some partnership practices as well. Agriculture here has always been about commodity production for sale, often to distant markets. It has always been fundamentally a business, and for that reason from its beginning has been conducted by "growers," not "farmers."[9] California's agriculture has been dominated by capitalist logic, and it has presaged trends in US agriculture.[10]

In California, technologically sophisticated and financially successful growers are termed "progressive growers." This is a term of respect, or even prestige, and it can also imply that grower is a social leader. Progressive growers are highly pragmatic and concern themselves with issues facing the agricultural industry. They see themselves environmentally responsible, and reject environmentalists' general criticism of their industry. Participation in partnerships allows them to demonstrate their claim that the agricultural industry cares about resource protection.

Twenty-four of the 32 partnerships have been in perennial crops (see figure 3.1). Perennial crops, perennial crop growers, and perennial crop organizations all appear to be more disposed to partnership activities for several reasons.[11] First, the perennial character of the crop is important. Perennial crops do not require the same degree of field disturbance that annual crops do (e.g., intensive tillage), and biocontrol strategies have been relatively more successful in them. The multi-year commitment of perennial crop agriculture means growers have invested in relatively few crops and cannot easily or cheaply switch to another cropping system. Second, perennial crops require a greater investment of capital and time to be successful. Even if the economics in another crop become more attractive, high transaction costs make replacement of an orchard crop unlikely in the short term. Stephen Welter quotes a saying: "growers in annual systems plow their mistakes under, while perennial growers learn to manage them." Perennial crop growers have to develop specialized knowledge to manage their operations successfully. The capital and knowledge investment appears to make perennial crop growers more disposed toward learning about their farming system. Third, perennial crop growers draw on a long history of cooperative organization, more so than growers who grow annual crops. Growers who produce high value commodities in spatially concentrated districts seem predisposed toward high levels of cooperation. Examples in perennial crop production include pears and winegrapes. Rice, strawberries, and artichokes are grown in concentrated regions and demonstrate a high degree of cooperation as well. Agroecological partnerships build on this tradition. Commodity organizations exist at the pleasure of a majority of growers, and have to continually prove their worth to their members to justify the assessments they collect from growers on each pound of crop.

Growers' Farm-Management Styles and Partnership Participation

Many researchers have described farmer (socio-economic) characteristics and the adoption of innovative techniques, following the "diffusion of innovation" tradition that started with rural sociological studies of hybrid corn adoption.[12] This approach has helped reinforce the "transfer of technology" extension pedagogy because it biases social-science research toward the novel technology, and away from the decision making process on the part of the grower. One particularly important exception is a recent study by Sonja Brodt and her colleagues investigating the role of farm-management style on grower learning about agroecological strategies and practices. Their studies of Lodi winegrape and almond growers demonstrated that management styles differ substantially among growers, that these differences affect how and why they access knowledge sources, and that growers selectively adopt knowledge about partnership practices consistent with their management style.[13] This work identified three distinct clusters of management styles: Environmental Stewards, Production Maximizers, and Social Networkers. This typology is fully consistent with my own depth interviews with participating growers. I will summarize their categories before adding my own observations.

Environmental Stewards place a higher priority on environmental stewardship than on getting the highest possible yields. Brodt and her colleagues describe them as wanting to "work in cooperation with nature and decreasing pesticide use as a way to improving living and working conditions." They are willing to invest time and money learning about new practices, and take some additional risks, because they believe they can improve their farming system. They want to make their farming system more environmentally friendly, and be able at least to break even over time, knowing that it will be perhaps less predictable. Many of them have taken some significant risks to innovate, and some became "leading growers" in partnerships.

Networking Entrepreneurs are oriented toward off-farm activities, especially those that allow them to learn about new techniques and technologies. Some of these activities were explicitly knowledge-oriented, such as work as a pest-control advisor or directing a local Farm Bureau. These off-farm activities expose them to information and technologies that Networking Entrepreneurs highly value. Farming is woven in with these activities, and they understand themselves to be better growers as a result. This group is highly attuned to watching other growers and how others perceive their own farming operation.

Production Maximizers see farming as a business, and they focus on producing the highest possible yield and quality. They focus on production and profitability, and eschew the distraction of off-farm activities. They express a competitive orientation, and a concern for the appearance of their farming operation. Some Production Maximizers portray farming as an industrial activity that, according to one respondent, was like "dueling with Mother Nature." They take a highly utilitarian view of natural resources, and regard farming as not harming the environment, but merely changing it.

By focusing on farm-management style, Brodt and her colleagues demonstrated that these three types of growers select different clusters of biologically based practices for different reasons, in contrast to the uniform and unidirectional assumptions of the transfer of technology and adoption/diffusion approaches. They suggest that while scientists and policy makers might wish every grower adopt an entire menu of practices, in reality a more effective extension strategy would batch practices into groups that correspond with different farm-management styles. That Environmental Stewards such as Glenn Anderson would be attracted to partnerships is unremarkable. The real significance of this study is their explanation of why other growers found agroecological partnerships attractive. More than 25 percent of Production Maximizers and 40 percent of Networking Entrepreneurs in the Brodt study participated in partnerships.

BIOS and the Lodi winegrape partnerships successfully cast their approach as new, innovative, socially enjoyable, and environmentally friendly—yet profitable. These appeal to the general orientation of producers toward innovation and pragmatism. Bob Bugg's depiction of the BIOS system as "charismatic" plays to these growers' concern with appearance. BIOS shifted grower perception of cover crops from being "weedy" to being attractive. Field days provided an opportunity for Production Maximizers to see how other growers were managing their farming operation and for Networking Entrepreneurs to exchange information with more people. Some partnership leaders have portrayed their efforts to have maximum social appeal while quietly pursing specific pollution reduction goals. To the extent they portray and can demonstrate partnership practices as profitable and innovative, they appeal to a broader audience of growers.

Many advocates of sustainability, including CAFF, have argued that smaller growers are more environmentally oriented, pointing to Glenn Anderson, who farms less than 40 acres yet supports his family. Environmental Stewards spanned the farm size spectrum, with several BIOS/BIFS participants managing more than 1,000 acres. Lodi winegrape grower Randy Lange fits the Environmental Steward profile, yet farms more than 6,000 acres. His business orientation, however, gives him credibility with other growers who do not share environmental values to the degree he does. Production Maximizers tended to have the largest operations, and Networking Entrepreneurs had the greatest percentage of smaller farms to facilitate off-farm employment.[14]

Participating growers are acutely aware of their geographic context: farming in a highly urbanized state, home to many people who see their industry as a threat to the environment. In US farming regions with rapid suburban sprawl, especially coastal states, growers are attracted to partnerships because they provide a vehicle to defend themselves against public antipathy toward farming. One vineyard manager in Monterey County said:

. . . I've been (a vineyard manager for) 30 years. For more than 20, I would say vineyards were considered the ecological friend of the state of California when it came to agricultural production endeavors. And somewhere in this growth period . . . we became an environmental concern.

Another winegrape grower in this region said:

I think I'm a good environmental citizen. Nobody's gonna know that unless I go out and tell them. And I need to tell them in a way that's irrefutable. . . . [This partnership] has the potential of . . . presenting a credible story and being able to back it up with good information.

Partnerships provide these growers a network to share information that allows them to learn how to farm in a way more consistent with their environmental values, but also a platform to present their farming in a positive light to their neighbors.

Many growers feel wronged by the way urban environmentalists have demonized agriculture. One Kern County almond grower and supporter of the almond PMA (although not formally enrolled) said:

I'm going to get on my soapbox in a minute. I think that farmers really get a bum rap for not being environmentalists, because basically that's what a farmer is. We live off the land, and we're environmentalists. The first thing we don't want to do is poison ourselves, our kids or our family, because we mostly live on the farms. And second of all, we don't want to sell a product to somebody and kill them with the product we sell, because then we can't resell the product. So I think we really get a bad rap that way.

This grower perceives an advantage in learning about agroecological pest-management strategies through the PMA because they represent a practical solution to the challenges of production, be they economic or regulatory. They see the advantages that agroecological approaches offer, and are happy to pursue them if they can be made to work. Another Kern County almond grower said:

We've gotten into this, I don't consider, neither [I nor my son] consider ourselves Greens by any sense of the imagination, but got into this type of pest management mostly because I feel it's the wave of the future. Regulations keep getting tougher and tougher. It's going to be harder and harder to spray.

This comment reflects a practical approach to staying in agriculture despite public opposition to pesticides expressed through regulations.

Growers participating in partnerships take a pragmatic view of regulations, and this appears to be part of the progressive grower profile.[15] They may have fundamental disagreements with environmentalists, but they accept the rationality of environmental regulations. A Kern County almond grower said:

My opinion of the environmental thing is: it's here. It's not going to go away. Deal with it. Work with it. It doesn't do any good to bitch and carry on and make a lot of noise about it. If you don't like it, go farm in Brazil. . . . They'll start whining and carrying on about the way things are, and I said "You know, you're in your fifties now. Why don't you just get out? I'm getting a little bit tired of listening to you bitch." [laughter] . . . And some do. Some finally get their fill of it, and for one reason or another, they just say "I'm tired of this shit, and I'm not going to take it anymore." (laughter) And that's fine. That's fine. I don't hold it against anybody for doing that.

This pragmatic view of environmental regulations was echoed by many partnership growers: they express some frustration and aggravation with regulations and complain about their costs, but they understand why they exist, and expressed a commitment to farming within them. For these growers, partnerships represent a practical way to be able to continue farming, and they actively support off-farm institutions, such as UC Cooperative Extension and their commodity organizations, leading partnership activities.

This section has summarized traits common to California's perennial crop growers and described the diverse motivations for wanting to experiment with agroecological practices. Individual growers have experimented with alternative practices, but partnerships extend the impact of this discovery learning through networks, but in order to do so with credibility, they require the participation of scientists.

California's Hierarchy of Agricultural Science

The work of Rachel Carson and Jim Hightower sparked public interest in the research activities of agricultural scientists and the consequences of their work. In 1983, Lawrence Busch and William Lacey published a groundbreaking book titled *Science, Agriculture, and the Politics of Research*, which was based on a rigorous survey of LGU scientists about their research activities.[16] Their work raised new questions about what constitutes "good" or "legitimate" science (and who has the privilege of determining these!), before the field of Science and Technology Studies was formally constituted. Busch and Lacey found that commodity groups were the most important external influence on LGU scientists' research agendas, and the selective financial and political support these groups offered shaped research trajectories. They also found clear evidence that these LGU scientists were ordered into a hierarchical system with basic research valued over extension practice, and that this system was on the brink of dramatic changes.[17] This section describes the different kinds of scientists and their respective roles—and motivations for participating—in agroecological partnerships.

California agroecological partnership leaders seeking scientific expertise immediately confront a scientific hierarchy shaped by institutional affiliation and function (figure 4.2). The most important fissure in this hierarchy is between research and extension activities. Research scientists have more prestige than extension scientists, and publicly funded scientists (supported by the University of California, and ultimately taxpayers) have more prestige than privately funded pest-control advisors. The University of California, like all land-grant universities, configures professional incentives and peer evaluations to clearly define roles for those who do research and those who conduct extension. Research scientists conduct basic research full time and sit atop this pyramid.[18] Immediately below them, Extension Specialists conduct basic research and some specialized applied research on a crop or a particular set of crops. The Extension Specialist position was originally created to link basic research with the applied needs of Farm Advisors and growers, although they now function essentially as researchers.[19]

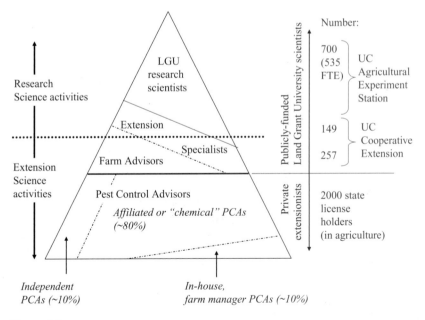

Figure 4.2
The agricultural science hierarchy in California.

Extension science activities in California are conducted by UC Cooperative Extension Farm Advisors and pest-control advisors. Over the past 20 years, UC leadership has restructured professional incentives for Farm Advisors. Located at the base of this scientific hierarchy, pest-control Advisors are the closest to the grower and the most economically practical in their orientation. They operate as agricultural science technicians, or subaltern scientists. Over the past few decades there has been a strong privatization trend in agricultural knowledge services, driven by several factors: increasing technological sophistication in agriculture, state legislative restrictions on pesticides, and a decline in public sector support for agricultural extension.

Agricultural Science in Action: Private Extensionists

Farm Advisors and pest-control advisors have developed a division of knowledge-labor in extension practice that allows Farm Advisors to retain certain privileges of scientific authority while PCAs conduct the majority of applied research, make all but a fraction of the pest-

---

California's Pest-Control Advisors

In California, only a pest-control advisor is permitted to make a written recommendation for the use of a registered pesticide material at a specific site. Between 80 percent and 90 percent of all PCAs work for agrochemical sales companies, and receive from them base pay plus a commission on all agrochemicals they sell. Critics refer to these as "chemical company PCAs," but they prefer to be known as "affiliated PCAs." Affiliated PCAs provide a seamless package of expert advice plus pesticide sales, delivery, and application, depending on the needs of the grower, but the "value" of the recommendation is folded in to the sale of agrochemicals. The balance of PCAs are either "independents," consultants who contract their services for monitoring and consulting directly with the grower, or "in-house PCAs" who hold a license but function as managers on corporate-owned farms. Independent PCAs are paid on a per acre basis. Those who have graduated from a University of California campus or have graduate training tend to be independent PCAs. Many report their status as independents having resulted in earning less money, but increased job satisfaction. They are exposed to more legal risk because as independent contractors they cannot afford "errors and omissions" risk insurance, which is provided by agrochemical companies for affiliated PCAs.

management recommendations, and earn more money. Full time growers and PCAs now use Farm Advisors only for extraordinary knowledge needs, such as novel pest identification questions. Many UC scientists speak of PCAs with disdain ("pesticide salesmen"), but are careful to point to the few exceptionally skilled independent consultants, who are able to develop applied management practices that more formal research-oriented scientists cannot.

PCAs occupy a unique position in California agriculture as consultants for growers who combine specialized university-generated expertise with practical, field-based experiential knowledge.[20] More than anyone else, they are able to observe regional trends in pest pressure and know the menu of tactics that can control them. Although there are only 2,000 of them working in agriculture, PCAs—after growers—are the most influential actor in pesticide-use decisions.[21] Even though California is the only state to license pest-control advisors, "certified crop consultants" fulfill similar functions in most farm states. In Washington, Karl Kupers helped start ACIRDS precisely because he could not find a consultant that had sufficient expertise to help him make the changes he wanted to in his farming system.

The notion of independent agricultural consultants pre-dates the state PCA licensing system. These early consultants pioneered the idea of charging growers on a per acre basis, wherein growers pay PCAs only for their expert knowledge. (For example, in almonds this fee is now typically $18–25 per acre per season, depending on what services beyond scouting are required.) Independent PCAs may operate more as scouts (where their exclusive responsibility is to monitor pest pressures), or they may be more complete consultants, offering specialized knowledge and writing legal pesticide recommendations.

Private consultants such as PCAs track regional trends in pest pressure and pesticide use. They have specialized training to observe and analyze specific threats that pests pose to crops, link university science with growers, and understand pesticide regulations. During the growing season they make regular visits to inspect crop health, monitor pests, and assess the risk of pest damage. They provide expert advice and help growers understand how to protect their crops. A PCA's specialized knowledge about pests, products, and regional trends is highly valued by growers.

Variability in the grower/PCA relationship thwarts general statements about how partnerships have impacted the pest-management decision-making process. Both affiliated and independent PCAs describe their role as largely dictated by the grower, and that they exist solely to serve their needs. In proposing a course of action, PCAs assess a grower's understanding of the ecological relationships in their field or orchard, as well as his or her financial resources, tolerance for damage, and risk aversion. A grower's approach to applying pesticides is shaped by these, but also equipment and labor availability, re-entry intervals for other activities (e.g., irrigation), and the anticipated cost of doing nothing. Experienced or sophisticated growers use PCAs like consultants; they are in charge, but they want an outside expert opinion. Growers that are smaller, without experience or confidence often "lean on their PCA" to help them make a decision. Smaller growers are less desirable for PCAs to serve consistently because they tend to demand a lot of time for consultation/education and "hand holding." PCAs will not receive sufficient compensation, either for pesticide sales or for per acre monitoring, to make serving them a high priority. Growers of this size are more susceptible to ad hoc pest-management decision-making. They are also more vulnerable to unscrupulous PCAs taking advantage of their ignorance.

In the first few seasons of a relationship, PCAs assess a grower's farm-management style, including his or her knowledge, financial resources, and attitudes toward hazardous materials. Many dimensions of grower-PCA negotiations are tacit (including potential unintended consequences). Thus, while advocates of reduced pesticide use criticize growers for "fear-based decisions," or PCAs "for preying on growers' fears," the risk of crop loss is so great that caution is economically rewarded. Growers know they will not lose their operation if they spray unnecessarily, but know that they may if they do not spray when required.

PCAs are quite frank about the risk dimension in their work. Affiliated and independent PCAs know that a wrong decision could push a marginal grower to the brink of economic disaster. Both independent and affiliated PCAs risk their livelihoods, and implicitly ask their growers to do the same, based on their expert knowledge. A PCA's continued occupation is dependent on his good reputation in the agricultural community, and if more than one bad decision is made, he or she may have to seek work in another field. Public extensionists are also likely to err on the side of caution and prescribe a pesticide. New PCAs have few incentives to reduce pesticide recommendations. They know that a bad reputation is much more difficult to repair than a questionable pesticide recommendation. The legal status of the PCA industry in California exposes PCAs to significantly more risk than field men in other states. Affiliated PCAs receive "errors and omissions" liability insurance through their employers, a legal protection that independent PCAs cannot afford.

Critics of affiliated PCAs claim they serve less of a link to UC knowledge and more as a filter, passing along information dependent on agrochemical use and screening out agroecological knowledge. Affiliated PCAs have been dogged by controversy—especially conflict of interest charges—since passage of the law that created them. Robert van den Bosch indicted them as members of the "pesticide mafia," but efforts to legislate a separation of pest-control recommendation and pesticide sales have not succeeded.[22] Conventional wisdom among partnership leaders says that independent PCAs promote IPM or strategies based more on agroecological knowledge than PCAs affiliated with chemical sales companies. This is often, but not always the case. Independent PCAs also are paid on a piecework basis, per acre instead of per pesticide. Critics of independent PCAs observe that they too prescribe pesticides in lieu of properly checking a field. Both affiliated and independent PCAs resist additional knowledge-intensive practices, perceiving them to be risky for growers and PCAs, although independent PCAs appear to have more confidence reading ecological data in the field.

Affiliated PCAs recognize the social and political risks of their industry's association with pesticides. They insist their industry is environmentally responsible, and express this forcefully, with talking points developed by the affiliated PCA trade organization, the California Agricultural Production Consultants Association. They acknowledge that selling pesticides is a controversial activity, but they insist they do so because they are agricultural professionals dedicated to the well-being of "our growers." They express concern that partnerships are promoting practices that ask them to do more and be compensated less, and they point out that these practices increase the risks posed to growers.

Partnership practices are welcomed if they entail selling products of equal value, but if the set of practices increases the amount of services and skills expected and reduces the potential for sales, it is less appealing. This industry manifests many traits of a quasi-profession or a "profession in process."[23]

Economic models of pest-management decisions that do not fully incorporate risk are inaccurate. Because of the biological uncertainty in an agricultural system, growers and PCAs never obtain all the information they would like, either across the farming system at the time of the decision, or predicting the future. Different pests pose different problems in uncertainty. Some pests, such as fungal diseases in almonds, have to be treated prophylactically, because once a fungal disease begins, fungicides cannot arrest it, only weather conditions. One former Farm Advisor turned grower described hoping to save $30/acre in fungicides, but he guessed wrong, and lost thousands of dollars of almond crop per acre that year and the next due to damaged bud wood. This illustrates the economic rationality of growers' reluctance to forgo fungicides.

Two factors emerge as antidotes to uncertainty and risk. The first is trust: growers who have established relationships with PCAs feel confident that their PCA is going to take care of them. It appears this sense of confidence helps growers manage their anxieties about the risks of farming. The second factor is knowledge, which agroecological partnerships promote through increased monitoring and knowledge of the farming system. Partnerships explicitly try to intervene in the farm-management decision-making process by making it more ecologically rational. PCAs have substantial influence over this process, and have been a secondary audience for most partnerships.

Partnership leaders report mixed success working with PCAs. They extol formally enrolled PCAs for their critical contributions toward making the techniques effective, but report that most affiliated PCAs express skepticism or passively resist pesticide reduction goals. Partnership leaders report being aware of the importance of PCAs to grower decision making, but many express concern that they do not fully understand how they influence growers. Only five partnerships designed their projects to enroll PCAs with their growers, and surprisingly few have conducted dedicated outreach to PCAs. Twenty-nine partnerships enrolled a total of 84 PCAs as formal partners, meaning that PCAs participated as applied consultants to the project, or were identified as partners. They have played critical roles in some partnerships, in terms of both the applied science and overall group leadership, although most of the time they prefer to operate in the background of partnerships. Established PCAs engage in partnership activity because it offers them the opportunity to sharpen their skills and develop professionally.

Enrolled PCAs have different skills and express their participation differently than do growers. Enrolled PCAs contributed less to almost every partnership activity than leading growers did, but their participation was nevertheless essential to partnerships' success because of their expert knowledge. Partnership leaders reported that PCAs were more important than growers in defining the problem that stimulated the creation of the partnership. Growers identified the problems and proposed a collaborative solution in more partnerships, but PCAs brought their regional perspective on pest management and farm management.

Partnerships promote agroecological strategies and practices among PCAs who are not formally enrolled. Many partnership leaders report encountering difficulties or challenges in trying to persuade PCAs to provide more sophisticated monitoring. Partnerships promote monitoring population trends for both pests and beneficial insects, recommending treatment only when damage is projected to be greater than the cost of pesticides plus their application costs plus the anticipated expense of additional pesticides required when the initial spray disrupts the existing biocontrol. Many of the partnerships have tried to help growers become better consumers of PCA services, such as by asking questions about thresholds, asking for written reports on monitoring, or asking about non-toxic, cultural controls.

Farm Advisors as Intermediaries

Farm Advisors formerly were a singular authority of applied scientific knowledge, serving a strong gatekeeper role. Private industry scientists and PCAs have expanded their own knowledge capacity to the point that UCCE prestige is relatively diminished. Farm Advisors still hold a privileged position by virtue of their historically tight relationships with University of California researchers, but private industry now accesses that knowledge and conducts its private research with relative ease.

Despite the antipathy of UC leadership toward the applied nature of partnerships, 92 UC personnel—scientists of all types—formally enrolled in partnerships, more than any other social group except growers. This represents roughly 8 percent of UC agricultural scientists.[24]

"Enrollment" for UC extensionists and researchers is formally designated by having their name (and reputation!) inscribed on partnership documentation, but it can have a range of operational meanings.[25] Farm Advisors perceive partnership as providing a service to growers by helping them manage regulatory uncertainty and the evolving social pressures on agriculture. This was one of the reasons Diana Roberts worked with the ACIRDS group. Several partnership leaders have worked intensively with the Farm Advisors who serve growers of their commodity. One of them reported mixed results: "UC Farm Advisors are the strongest aspect of what I have in my program, compared to other programs. It [sic] is also the most difficult problem to deal with in my program. Even more so than the growers." This comment reveals how difficult it can be to promote agroecological perspectives among Farm Advisors. A few of them see agroecological partnerships as a way to better serve growers, but many also perceive them as yet another intrusion by yet another institution into their work. Several observers have noted that an individual Farm Advisor may quietly contribute critical knowledge resources to a partnership, but may also exert pressure on their peers and publicly speak against further erosion of UCCE scientific authority.

One specific kind of Farm Advisor has taken to partnerships enthusiastically, the UC IPM Advisors. When the California Legislature created the UC IPM program in 1979, it included six regional IPM Advisors who would promote field implementation of IPM practices in clusters of similar crops.[26] Not surprisingly, IPM Advisors had the highest proportion of personnel participating in partnerships: five of the seven contributed to them. Two IPM Advisors, Walt Bentley and Carolyn Pickel, participated in eight partnerships each, more than any other individual. Their standing professional responsibilities are research and grower education on IPM in tree crops. Upon their formation, many partnerships consulted with IPM Advisors because they had much of the necessary applied knowledge. Partnership leaders frequently lauded IPM Advisors for their singular contributions. Partnerships were a boon to IPM

Advisors because they provided a professional and funded forum to promote IPM techniques, and they dedicated large portions of their time to partnership activities. Pickel said: "I have been working in this area for many years. I like the PMA [partnerships], because it's like, someone started paying attention. You're not carrying the banner by yourself, or with the other Farm Advisor, beating your head on the wall. It's sort of like the industry said, we need this. And it's supporting groups of people to get implementation. To me, it's my job." The reason no IPM Advisors were principal investigators on grants was that they have multi-county, multi-crop responsibilities, and they were used primarily as expert knowledge resources by multiple partnerships. Current job performance criteria for Extension Specialists orient them toward basic research, which means that they have fewer incentives to conduct extension activities. Unlike the IPM Farm Advisors, Extension Specialists did not play a prominent supporting role; they were Principal Investigators, or they played a minor role, limited to providing technical advice.[27]

Engaging Agricultural Researchers

Securing the expertise of UC researchers has been one of the more difficult challenges facing partnership leaders. Precious few of the 700 agricultural research scientists contribute to alternative, agroecological strategies. Despite the public charter and public financial support of LGU science, partnership leaders describe difficulties in recruiting UC research scientists to work on practical problems. The President of the Almond Board visited the UC Davis campus and reported UC to be in retreat from production agriculture, and characterized this as a "progressive disease."[28]

All 32 partnerships relied on UC research science, albeit to varying degrees.[29] Most leaders designed their partnerships to take maximum advantage of this knowledge by extending it more fully through outreach activities. The proportion of dedicated research scientists participating was tiny—only 2 percent—but they have had a huge impact on partnerships because their scientific knowledge and skills have been pivotal to the success of many agroecological partnerships.[30] For example, Dwayne Beck of the Dakota Lakes farm in South Dakota played an essential role in the impact of the ACIRDS/Wilke Farm partnership more than 1,500 miles away.

The few LGU research scientists who have contributed to partnerships report that participation brings them personal satisfaction for several reasons, despite the professional and institutional disincentives. First, and probably most important, partnerships provide them a group of relatively open-minded growers who are willing to try risky, innovative practices that improve environmental protection. These farmers have the skills and disposition to successfully learn and implement sophisticated experimental practices, and have an above average tolerance for risk. The importance to these scientists of finding appropriate cooperative growers should not be underestimated. Thus, growers and their organizations enroll scientists in an agroecological initiative, while the scientists simultaneously enroll the growers in his or her environmental science research project. Second, some scientists like taking on an applied professional challenge. Based on his success in disrupting the codling moth in pears, Stephen Welter was successfully wooed by the Walnut Board to work with them, and he wanted to see if he could extend this success further. He was intrigued by the difficulties of adapting an approach that worked in pears to the more challenging walnut farming system. Another suggested that scientists want to work on systems that are not well understood; this work is more interesting, and perhaps more scientifically valuable. Third, partnerships offer them an opportunity to contribute to efforts to reduce pesticides. "Reducing pesticide use," one scientist said, "gets me up in the morning." Many contemporary agricultural scientists entered their profession with environmental values, but find that the demands of their position do not easily lend themselves to working on pesticide reduction. In addition, commodity organization sponsored projects have the ability to reach every grower in the state with newsletters and bulletins explaining low input practices. Their personal orientation and professional training provide them unique skills, but they must step out of their designated roles within the science hierarchy to put them to use in partnerships.

LGU scientists report that practical, applied work in agriculture can have a detrimental effect on their professional reputation within the academy, and they have to carefully manage their scientific reputations. Welter counsels his graduate students to defer their interest in practical research: "I don't recommend it for my students, because 'they can't afford it' yet. And if they choose to do that on their own, then that's their business. . . . What I say is, as a graduate student, you need to set up your credentials in an academic environment, and then you can expand your interests. But I think that it's very risky as a graduate student." Partnership activities are a form of moonlighting for most LGU research scientists. They offer scientists an opportunity to interact with skilled growers and PCAs and contribute substantively to the well-being of agriculture, but they have to be careful it does not interfere with their scientific careers.

New Roles for Agricultural Organizations

Partnerships build on existing social networks in agriculture. Commodity-specific organizations have played crucial facilitating roles in agroecological partnerships. They pursue the interests of their member growers (at least the larger and influential ones), while helping growers recognize that the social and regulatory context of agriculture has changed since *Silent Spring*. Other types of grower organizations include non-governmental organizations, grower cooperatives, trade organizations, and informal groups like ACIRDS. For the purposes of partnership activities, the legal status and original purpose of a growers' group is less important than the development of environmental

The Commodity Organization

Early California growers and the state created commodity-specific institutions to address marketing and production research. These were historical antecedents to the commodity organizations participating in agroecological partnerships starting in the 1990s. Commodity organizations originally emerged to help growers manage the cross-country transportation of their fruit but their roles expanded to help growers in all phases of their industrial development. Growers were able to lobby successfully for enhanced status and authority for them during the farm crisis of the early twentieth century. In 1915, the California legislature passed the first laws enabling fruit marketing cooperatives. In 1937, the California legislature and the US Congress passed laws based on these cooperatives to help producers cope with Depression-era prices. There are now 33 federal and 41 state crop commodity organizations. The single-crop focus of these organizations tends to reinforce commodity-specific knowledge systems, and predisposes them to play a facilitating role in agroecological initiatives.

leadership within it. The entrance of these kinds of organizations into extension activities marks a prominent, novel development in American agriculture, and distinguishes the agroecological partnership model from traditional extension.

Commodity-specific organizations can be highly influential among their growers, many of whom look to them for information about practices and regulations. Federal and state commodity organizations are semi-public corporate groups financed by a mandatory tax imposed by a majority vote of the growers of a specific geographic area. All but a few have the primary purpose of promoting the marketing of a particular agricultural product through promotion, advertising and the imposition of quality standards. Most are also authorized to fund production research, although usually only a tiny fraction of organizational budgets are devoted to this. I use "commodity organization" as a general term for commodity-specific orders, councils, and commissions.[31]

Not all commodity organizations are equally interested in partnerships. Perennial crop commodity organizations manifest a belief that it is in their best interest to provide leadership on pest-management issues, while organizations representing growers of federally subsidized crops (e.g., corn, cotton, peanuts) have generally adopted a passive stances or in some cases, actively resisted agroecological initiatives.[32] Perennial crop commodity organizations have invested significantly more resources in helping growers find alternatives and negotiating with environmental regulatory agencies. They are motivated by their concerns about environmental regulations and in some cases, the desire to represent themselves and their growers as environmentally responsible. Much of the responsibility for coordinating partnership activities falls on the desk of the commodity board's research director. They generally express enthusiasm for this new role, but it entails a considerable amount of extra work. Research directors often play crucial roles within their respective commodity. They broker knowledge and financial resources between regulatory staff, researchers, and commodity growers.[33]

The Almond Board of California has dedicated more effort to environmental issues than any other non-winegrape commodity organization. The ABC is the only commodity board to create an environmental committee. In the words of Ray Eck, an early organic BIOS grower active on this committee, "the consciousness is that the industry has to gain by cooperating with these kinds of challenges." This is a novel development in the history of commodity organizations in California, which have traditionally limited their focus to "flow to market," promotion, and production research. Like most human undertakings, the Environmental Committee was created from more than one motivation. It was created partly out of a defensive posture, but also out of a desire to negotiate enforcement of environmental regulations.[34]

Chris Heintz, the director of production research and environmental affairs at the ABC, describes her negotiations with regulators as a new dimension of her position, one that takes an increasing amount of her time. Her associate Mark Looker said: " [The ABC is] definitely going to do all we can to inform growers and PCAs and regulatory people and elected officials about the fact that we are doing something that is looking in to what the problem is. I don't know if that is a philosophical difference between different commodities. We have always been very open about the fact that we have problems and we'll try to deal with them, and we don't have all the answers, but we definitely think that you have to have a science-based approach to finding solutions." Heintz observed that when she took her position in 1996 she never expected that she would be coaching growers how to speak to regulators, and vice versa. She does not feel that negotiating with regulatory agencies changes the regulations, but it does allow the industry input to how the regulations are implemented.

Reaching out to growers accorded the Almond Board credibility with public agencies because it effected change in a way a regulatory agency could not. The ABC once invited USEPA staff on a tour of almond orchards and the processing industry, not unlike the "listening tour" on the Columbia Plateau. According to Looker: "It is so much better to say 'we as an industry are taking these approaches and our partners are CAFF, the Almond Hullers and Processors.' . . . It just gives us so much more strength to talk to them. They realize you are bringing all the players to the table, and everyone is having input into these issues, I think it gives you a lot stronger foundation, especially with the regulatory folks." Thus, the ABC inserted itself as a buffer between growers and regulatory agencies, strengthening its standing. It has been able to present itself to growers as their friend, keeping regulators at bay and proposing voluntary yet viable alternative pest-control strategies. At the same time, it has been able to present itself to regulatory agencies as rational, responsible, and demonstrating leadership in urging its growers to change. Commodity organizations were by far the most important institutional participants in California's partnerships, and the PMA program explicitly tried to capture their interest. They were the lead organization in eight partnerships, writing grants, recruiting partners, and helping design the research and outreach.

Other kinds of grower organizations participated in smaller scale partnerships. Two "grower cooperatives" and two trade organizations have been involved in partnerships. The Sun-Maid Raisin Growers of California is an example of the former, and the Sonoma County Grape Growers Association of the latter. Because raisins are dried and stored for up to 2 years prior to consumption, the Sun-Maid Raisin Growers felt particularly vulnerable to pesticide-related food safety issues, and helped their growers reduce reliance on pesticides.[35] The Sonoma County Grape Growers Association launched a partnership to address local concerns about the environmental impact of its growers' practices. In general, trade organizations engage in partnerships as a way to help growers cope with pesticide-related concerns. CAFF was the dominant NGO partner, participating in eight. It was the lead organization in three, and coordinated grower outreach for the other five. Growers and others have created several other nonprofit organizations to assist them with partnership activities.[36] These organizations provided staff to coordinate partnership activities and facilitated the reception of grants. In some cases, growers have affiliated informally, without an organization name or a legal identity.

Environmental Agencies and Partnerships

Several approaches on the part of environmental agencies emerged in the 1990s as they began recognizing that traditional regulatory efforts did not adequately address agricultural pollution issues, especially non-point-source water pollution. CalEPA/DPR and several regional offices of the USEPA have provided funding, leadership, and technical assistance to agroecological partnership activities. USEPA staff member Augie Feder was able to use the language of pollution prevention to persuade the San Francisco (Region IX) office to fund the Biologically Integrated

Orchard System (BIOS) and BIFS. Jean Marie Peltier used the discursive image of "public/private partnerships" at the DPR to secure funding for the PMA program. Chris Feise and Seattle office USEPA (Region X) staff argued successfully to their supervisors that the Columbia Plateau Agricultural Initiative proposal would fulfill the integrated environmental goals (i.e., air and water) of the USEPA Regional Geographic Initiative program.[37] These emerged as agency personnel recognized the limitations of the "command and control" approach to environmental regulations and sought more community-based, multi-disciplinary, and integrated approaches.

The conflictual relationship between the USEPA and the industries it regulates can be traced back to its inception and first administrator, William Ruckelshaus. He was a prosecutor who had been quite successful in addressing water and air pollution through the courts in Indiana, and his experience convinced him of the merit of unambiguous standard setting and aggressive law enforcement. Negotiation and assistance were not in the USEPA's policy tool kit for its first 20 years of existence. Ruckelshaus understood the USEPA as a "gorilla in the closet," a potent legal force that could be brought forth to inflict legal and financial pain on an industry if it was not cooperative with local regulators.[38]

The DPR and the USEPA are chiefly regulatory agencies, but they developed new roles in agriculture by participating in partnerships. The entry of regulatory agencies into extension activities provokes ambivalent feelings among agency personnel and in the agricultural community. Agency personnel have asserted the primary function of their agency is to enforce environmental laws, even as they provide to some organizations of growers who are in violation of the Clean Water Act. Agricultural organizations welcome the funding regulatory agencies' dollars, but have had to reassure growers that they are not "negotiating with the enemy." Agency personnel welcome the opportunity to be seen as playing a constructive role in the agricultural community, and agricultural organizations represent themselves to regulatory agencies as being environmentally responsible. Eight other public funding institutions have participated in partnerships, mostly by providing small grants.[39]

Despite feelings of ambivalence held by agroecological partnership participants, a recent analysis of the CPAI initiative determined that growers, the USEPA, and other public sector participants believe it was successful, and an important first step into a new way of thinking and engaging in community-based pollution-prevention initiatives. USEPA staff perceive the agroecological partnership approach has many limitations, but all the other "tools in their toolbox" are even less effective. Both regulators and growers observed that the "listening tour" may have been the most significant strategy because it laid the foundation for collaboration. To replicate some of the success of CPAI will require committed staff members, growers, and community members dedicated to local leadership, and adequate long-term funding.[40]

Conclusion

Partnerships emerge to address unmet needs and to progress toward goals that cannot be achieved by individuals. An agroecosystem is a functional system of complimentary relations between living organisms and their environment. Agroecological partnerships organize participants to create synergistic benefits from social learning interactions. To progress toward partnership goals, actors develop new roles and recognize the potential benefits of shifting existing relationships.

Latour's circulatory system of science brings order to the many different kinds of participants, and helps explain their institutional location and motivations for making contributions. All four external loops represent major categories of partnership actors, and they circulate knowledge to achieve their own needs, as well as contributing to the good of the collective. Growers learn to perceive the benefits of bringing additional perspectives on their farming system, and to incorporate new insights into their farm management. Extensionists learn to recognize the benefits of serving growers' needs as growers respond to social pressure for improved stewardship. Even as science institutions evolve away from meeting practical needs in agriculture, some researchers welcome the opportunity to contribute their expertise for social benefit. Grower organizations have developed new roles in representing their grower clients as they respond to increased regulatory pressure. Public agencies are learning the value of collaboration with agriculture, even as staff members struggle with ambivalent feelings. The next chapter describes the kind of knowledge these partnership participants generate and circulate.

5

The Practices

The Alluring Almond

Plants do amazing things with chemicals. Fixed in place, plants use them to repel or attract their ecological companions. The wild "bitter almond" (*Prunus dulcis* var. *amara*) was so named because it contains amygdalin, an intensely tart chemical compound that breaks down to cyanide. It evolved in the mountains of west-central Asia, where winters were warm and wet and summers hot and dry. Its chemical defenses repelled those companions that were tempted to sample its seeds, and its poison punished those that disregarded its bitterness. The presence of amygdalin is controlled by a single gene, however, and occasionally a mutation would sprout. Some early farmers identified certain mutations and began to cultivate them. Evidence of wild almonds can be found at Greek archeological sites dating back 10,000 years.

Ancient peoples moved the domesticated almond (*Prunus dulcis*) around the Mediterranean Basin, where they learned by trial and error to plant it on rocky hillsides so as to avoid early spring frost, which destroys its flowers. Almond orchards were usually marginal: they occupied poor soils, were generally dry-farmed, and added a little extra to farmer families' nutritional needs and income. The early Franciscans brought almonds to several California Missions. The first commercial almond orchard in the state was planted near Sacramento in 1843 with varieties from the Languedoc region of France, but commercial production eluded growers for decades. Growers did not understand the need for bees to cross-pollinate different almond varietals, and they did not adequately protect their trees from frost and disease. Thus, almonds came relatively late to commercial orchard production in California, after all manner of citrus, stone, and pome fruit.

In 1879, A. T. Hatch of Suisun planted some 2,000 seedlings to graft, but ran out of budding material for the last 200. Several of these ungrafted seedlings performed particularly well together, including one he named Nonpareil, which became California's most popular and valuable almond variety. By happy chance, three other good varieties flowered concurrently with the Nonpareil, providing all the necessary pollen. California almond growers are still completely dependent on bees to pollinate their crops. They rent hundreds of thousands of beehives from Arizona every winter to ensure they have enough. Almond growers continue to support research on bee behavior through their commodity board, and might be more attuned to ecological relationships in their farming systems than other growers.

Of all the insects active in almond orchards, only five species cause economic damage. The two chief pests today are both lepidopteran. The peach twig borer (*Anarisia lineatella*) is native to Europe, where it attacked stone fruit and almonds. Carried to California in the 1880s, it was the major almond pest until the 1940s, when the navel orange worm (*Amyolois transitella*) arrived. The lifecycle of the navel orange worm (commonly abbreviated to NOW) meshes tightly with almond development. In almonds, the NOW found a remarkable ecological companion that offers it food and shelter. Nourished by water and agrochemicals, almond orchards proliferated throughout the Central Valley after 1960, but these brought biological changes that made them more hospitable to the insect. In the late 1970s, NOW crop damage reached a crisis which provoked growers to reconsider their chemical pest-management strategies.

Some controversy surrounds the navel orange worm's precise evolutionary locus of origin. The UC IPM manual for almonds describes it as a native of the southwestern United States and Mexico, but Fred Legner and others believe that it originated further south in the Neotropics, perhaps in Argentina. Navel orange worms mature into gray-and-black moths about 12 millimeters long. During the cool spring weather, adults live 2 or 3 weeks. During hot summers, they survive less than a week. Adult females mate within two days of emergence from their larval stage and begin to lay an average of 85 eggs each. These eggs are less than 0.25 millimeter across. They hatch in from 4 to 23 days. The larvae then forage for several days or weeks before emerging as adult moths to reproduce. The navel orange worm's progress through its life stages, like that of most insects, is conditioned by environmental heat units. There are three or four flights (generations) in a year. NOW do not diapause (hibernate) during the winter and must have continual access to food.[1]

The navel orange worm migrated to the Central Valley just as almonds were being established as a commercial crop and as this crop's center of production shifted from the coastal counties to the Sacramento Valley. Initially almonds were dry-farmed, and the harvest labor requirements constrained its expansion. Workers knocked almonds from the trees with mallets and poles, passing through the orchard several times as varieties ripened, swept the nuts onto tarps, and dragged them to the orchard edges by horses. This was done in late summer, when all growers were competing for labor.

In the 1950s, significant investment by the state of California in the University of California's agricultural research and massive new water projects revolutionized almond production. UC contributed a package of improved production techniques. Mechanization reduced labor requirements by 75 percent. New, shorter tree varieties were more productive. Chemical fertility and pest-management programs boosted yields. The "shaker"—looking like an enclosed, low-hung go-kart—zips from tree to tree, seizing the trunk with a huge mechanical arm and shaking the nuts off. This device made it economically efficient to harvest an entire orchard block at once. A mechanical sweeper follows, shaping a long, continuous nut pile down the row middles. Finally, a loader scoops up the nuts which conveys them to a semi-truck at the edge of the orchard and transports them to the processor. The State Water Project brought irrigation to hundreds of thousands of new acres in the southern San Joaquin Valley, provoking a tremendous expansion of almond acreage. With support from the Almond Board, the insecticides azinphosmethyl and carbaryl were tested and registered for use on the navel orange worm in 1976. In the market, almonds evolved from a rare treat to a specialty crop, and are now a common industrial input for snack foods. Both humans and insects find them tasty.

Revolutionary practices have re-configured the geography and ecology of almond production. California growers now produce 80 percent of the global almond crop. Yields per acre climbed from an average of 213 pounds per acre in the 1920s to 1,348 pounds per acre in the 1990s. Almonds are currently the most extensively planted tree crop in the state, and account for nearly 5 percent of California's irrigated cropland. The labor efficiencies achieved by harvest machinery have made almonds an attractive crop for small and medium-size growers.[2]

Increased production did not come without costs. Almonds develop inside a shell enclosed by a hull. This double jacket protects the nuts but is not impermeable, and it can offer shelter to the pest. Chemical nitrogen plus irrigation boosted tree growth and increased nut size, making vegetation and the developing nut more attractive to pests. Almond hulls are analogous to fruit flesh of their plant cousin, the peach. In the late summer, a few weeks prior to harvest, hullsplit occurs, meaning that the hull peels back to reveal the nutshell. Simultaneously the fibers holding the nut to the stem begin to degrade. Trees require not too much and not too little moisture for hullsplit to proceed properly. Overly vigorous trees will delay hullsplit, and the trees stressed by inadequate moisture misfire the ripening sequence, with hulls failing to release the shelled nut.

Larvae of the navel orange worm and the peach twig borer sniff out the developing almond oil and try to penetrate the sealed hull throughout the spring. Several of the new, high production cultivars have softer shells, meaning that peach twig borer larvae chewing through the hull easily work their way straight through to the nut. During his 40-year career, UC Farm Advisor Lonnie Hendricks watched the NOW become more of a pest as almonds became a widely planted and highly fertilized monocrop, in part because the shell and hull seal began to open more readily. As almond acreage grew, so did NOW food and habitat. The NOW is a non-native insect, and thus has no native natural enemies to dampen explosive population growth. As the 1960s drew to a close, the NOW surpassed the peach twig borer as the chief economic pest in almonds.

More and more pesticides were applied, but to little effect. The typical pest-management program in the 1970s consisted of "holiday sprays" by the calendar. New Year, Memorial Day, and July 4 roughly approximate the times NOW larvae moved about, but variations by year in heat units could make those sprays worthless. Nut damage rates due to the NOW began to climb from the single digits up to 20, 30, even 40 percent (figure 5.1).

Figure 5.1
An almond that has been ruined by the feeding of a NOW larva.

Earl Decker grew up working in his father's almond orchard in Durham, near Chico, during the 1940s. His father had a long history of cooperating with Farm Advisors in research, and this orientation toward scientific solutions was apparently passed on to his son. In 1978, Decker's NOW damage rates were so high as to make him pull his hair out. He brought pest-control advisors, extension agents, and several scientists to his orchard, but none were able to control the NOW. Decker realized they needed more help, so they contacted George Post and Bob Hanke, the first independent PCAs in the Sacramento Valley. Post and Hanke replied that they would be willing to help them with their problem, but they would require $35,000 as a down payment. Decker and four other growers pledged the full amount that same day, and organized a meeting at Decker's house. Post and Hanke came to investigate, and then contacted George Okamura, a scientist recently retired from the California Department of Food and Agriculture, who had managed the Los Angeles Medfly crisis. Okamura determined that the NOW overwintered in "mummies" (un-harvested nuts that did not drop from branches), which perpetuated large populations the subsequent season.

Changes in the behavior of almond trees had made them more hospitable to the navel orange worm. Growers stimulated almond production, and an unintended result was increased numbers of mummies. Once a crawler (larva, commonly referred to as a "worm") found shelter in a mummy, it was home free, with old nutmeat for food and near-complete protection from pesticides. The first generation of adult females in the early spring would seek out mummies, attracted by almond oil, and lay eggs nearby. This guaranteed that high numbers of NOW would hatch in May, which in turn would reproduce and flood orchards with their third generation, assuring high nut damage during the vulnerable hullsplit stage. Subsequent research would reveal that the peach twig borer and the NOW work as a tag team: peach twig borer larvae chew through newly forming green hulls and shells, and the subsequent generations of NOW can then access the nutmeat. Thus, the peach twig borer creates a "food bridge" allowing NOW populations to build up in the spring before hullsplit, and multiply rapidly once all nuts become vulnerable.

Okamura, Post, and Hanke explained the life cycle of this pest to the growers. Organophosphates could be effective, but only if they were sprayed precisely when crawlers were seeking out the nut. Spraying at the wrong time would make matters worse by killing off any generalist predators and causing an explosion of mites. But they emphasized that orchard sanitation in early was the most important technique: shaking the trees to knock off mummies, and flail mowing them on the orchard floor. Mummy removal must leave no more than one nut per tree. Sanitizing one's own orchard is effective, but the effectiveness is increased if it is done on an area-wide basis since NOW adults can fly up to half a kilometer.

Bob Hanke did not think the growers would be willing to sanitize, but they were. Decker described a palpable crisis among growers that provoked interest in social learning beyond the norm. The leadership of certain growers was very important to the success of this initiative. They identified the problem, self-organized, invested their own resources, and recruited specialists to address it. Individualism is the norm among growers, but the NOW crisis forced this group to develop new social relations. When Okamura asked the growers who sponsored him if they would like to share the information with others, one of them, well known for keeping to himself, said "I don't mind sharing anything with anyone else as long as they want to kill worms!"

Growers and PCAs in Butte County point to this social learning process in the late 1970s as convincing them that sanitation is more important to NOW control than pesticides, but UC scientists had identified cultural strategies as early as the 1950s.[3] Scientists frequently conduct research in advance of problems, but its movement into the field is filtered and warped by the social processes of extension, both public and private. UC scientists and extensionists have to simultaneously pursue fundamental scientific research and provide information meaningful to PCAs and growers, who are reluctant to alter their practices as long as they are profitable. Before the Durham growers ponied up funds to pay for independent PCAs to determine the cause of their NOW problems, the US Department of Agriculture and the Almond Board had already received results from research into sanitation and early harvest they had funded. Charles Curtis of the USDA and Martin Barnes at UC Riverside determined that leaving one nut per tree could cut an orchard's NOW population in half. Curtis was inspired by Rachel Carson and the IPM pioneers. The USDA funded the "Ballico/Famoso Project" for more than $435,000 between 1974 and 1976 to investigate the benefits of regional almond IPM practices.[4]

In 1980, the Almond Board and the new UC IPM Project initiated a four point program for NOW control: dormant spray, orchard sanitation, properly timed in-season sprays, and early harvest. Over the past 20 years, California's almond growers and PCAs have learned that cultural practices are equally or more important than pesticides in controlling NOW populations. The validation of the UC IPM program has been critical to the almond industry's success. Even affiliated PCAs acknowledge that sanitation is the key to controlling the navel orange worm, although they gladly sell organophosphate pesticides for in-season use to compliment cultural controls.

The almond industry's social learning about NOW control is akin to the citrus industry learning about biocontrol in the nineteenth century. The almond industry in the 1980s was recognized as an early IPM success story because it relied heavily on ecologically rational, non-chemical strategies to control its primary insect pest, even though many UC Farm Advisors, growers, and PCAs were reluctant to rely exclusively on ecologically beneficial organisms as an agroecological pest-management strategy.[5] The two almond partnerships subsequently demonstrated that organophosphates are rarely needed as dormant sprays.

The BIOS pioneer Lonnie Hendricks's professional career as a UC Farm Advisor coincided with the explosion of almond acreage and the growing awareness of IPM in California agriculture. In the 1960s he helped the UC scientist Leo Caltagirone establish the parasitic wasp *Copidosomopsis* (or *Pentalitomastix*) *pletheorica*, which provided some, but not consistent, control in unsprayed orchards. In the 1970s, Hendricks participated in the Ballico/Famoso Project, which took place among a cooperative organization of growers in Merced County. The Cortez Growers Association was able to coordinate area-wide sanitation effectively, and they still cooperatively hire one independent PCA to monitor regional pest populations. In 1980, Hendricks helped the UC Riverside entomologist Fred Legner release *Goniozus legneri*, a parasitoid from Argentina and Uruguay.[6]

The *G. legneri* wasp crawls up the back of a NOW larva and stings it at the base of its head, injecting a toxin (figure 5.2). The NOW thrashes about, trying to dislodge the parasite, but to no end; the venom induces paralysis. The NOW's heart continues to provide blood to the body, but it cannot move. The wasp lays up to a dozen eggs on the larva, depending on its size. In about two days, the eggs hatch and the wasp larvae begin to feed on the NOW's body. They consume the larva, leaving only a head capsule and fragment of skin, and then pupate and emerge as adults. Their life cycle completes in 15–20 days. Fred Legner determined that *G. legneri* remains active through the winter, depending on the NOW for food. He concluded that sanitation was "certainly counterproductive to maintaining parasitoids of navel orangeworm," and observed that research supportive of sanitation was conducted prior to the widespread release of the parasite.[7]

When the BIOS pioneer Glenn Anderson asked his Farm Advisor to investigate how he had successfully grown almonds without insecticides, Hendricks took him seriously because he had witnessed biocontrol in action. When he reported his pre-BIOS research to SAREP in the late 1980s, Hendricks wrote in boldface: "**Nonpareil variety can be grown without pesticides.**" Anderson recalls: ". . . it was previously assumed that the nonpareil, because its seal is ruptured, is incredibly susceptible

Figure 5.2
Top: A *Goniozus legerni* crawls up the back of a NOW larva. Middle: The *Goniozus* paralyzes the NOW and lays eggs on the back of its neck. Bottom: *Goniozus* eggs emerging from the NOW carcass. Photographs courtesy of Kent Daane.

to insect damage. It was totally assumed the nonpareils could not be grown without pesticides. [Hendricks's report said] 'It has become the least of our problems. We have low rejects on nonpareils as we do other varieties. The seal is not much of a factor.' And he stated that boldly in the middle of his report. I thought, 'Whoa. Now we're making headway.'" Thus, before BIOS was even conceived, the scientific inquiry of a critical partnership pioneer led him to question some dominant assumptions about almond pest management, in part because he had a lived experience, shared with other scientists, of studying nature to make once-hidden ecological relationships visible. After Hendricks started to report his findings, he encountered controversy, according to Anderson: "I went to some of his colleagues' meetings in other parts of the valley. I mean they were roasting sessions. 'You're out of it, Lonnie. You're going to be back. This whole thing you're doing is taking you the wrong direction. You're going to be in trouble here. You cannot be even suggesting that this stuff is right, that this could be the right direction.' Oh yeah. I went to a couple of those with him." Other Farm Advisors observed professional courtesy in public with Hendricks, but not with CAFF staff once they began an outreach program. Affiliated PCAs were none too happy with him either. Hendricks's experience with biological control, the strength of his scientific training, and his confidence in observation convinced him BIOS could work. Because of this, he was a true scientist, according to Anderson. He did not say growers had to follow BIOS principles, but he was sure they could succeed if they wanted to.

Efforts to put agroecology into action have confronted a public and private extension system that depends on a science of universal truths, when in fact pests and beneficial organisms are dynamic in time and space. Temperatures and precipitation vary just enough along a north-south gradient in the Central Valley, and cool summer breezes across the Sacramento Delta moderate temperatures slightly in the northern San Joaquin Valley. Almonds and navel orange worms do not develop at the same pace in Butte and Kern Counties. *G. legneri* does not consistently survive winters north of Merced County, so it must be augmented from an insectary. A 2001 study led by Walt Bentley failed to find consistent control by *G. legneri*, although many BIOS growers and several independent PCAs find it effective.[8] Broad spectrum insecticides mask the ecological variability of a farming system. Taking advantage of agroecological strategies requires an approach to decision making that can account for geographic variation and ecological dynamics.

When non-UC extensionists involved in the BIOS and Lodi winegrape partnerships promoted alternatives to the dominant UC recommendations, they often encountered controversy even when UC researchers themselves had varied views. For example, the BIOS manual conveyed Legner's suggestion that removing mummies could deny overwintering habitat necessary for his namesake parasitoid, making caveats that this could increase NOW habitat as well, and recommending growers develop their own pest-management strategy whole plant/insect system. In practice, some BIOS staff did not effectively communicate the full range of risks, or facilitated field days that failed to do so. Some Farm Advisors attacked BIOS for promoting "anecdotal" information. BIOS publications promoted UC-generated agroecological knowledge on such things as cover crops, monitoring, early harvest, and sanitation, but in the eyes of many Farm Advisors this could not redeem the partnership from the sin of presenting information that was not universally true.

Agroecology in Agricultural Partnerships

Putting agroecology into action begins with mobilizing organisms—including humans—into new patterns of relationships guided by ecological knowledge in particular places. For agroecological strategies to succeed, participants first have to be motivated to seek alternatives to chemical-intensive farming. They demand different types of knowledge, ranging from the theoretical to the practical. These kinds of knowledge depend on the kind of social learning narrated above.

Farming systems are not static, but rather dynamic across time and space, and to manage them with agroecological strategies and practices require an adaptive approach.[9] Agroecological learning binds people more closely to nature and its ecological organisms and relationships. The careful observation and monitoring necessary to manage these organisms and relationships is labor intensive. Growers and scientists have to invest more in gathering site-specific and time-specific agroecological knowledge. Agroecological initiatives seek to replace the potency of technology, chiefly agrochemicals, with the vitality of living organisms. Partnerships conceptualize bio-diversification and its synergistic benefits within the context of economic monoculture. Bio-diversification in California means the introducing and management of the beneficial organisms, chiefly cover crops and natural enemies, although in other initiatives, such as intensive rotational grazing, agrobiodiversity is managed in other ways (e.g., pasture plants). As growers remove ecologically disruptive and hazardous organophosphates from farming systems, they are presented with new opportunities to manage ecological organisms and relationships. This approach can be effective, but is generally less predictable than manufactured technologies. Few of the growers, the

PCAs, or the Farm Advisors I interviewed used the term "ecology" or "agroecology," but they made frequent reference to how they have come to perceive and manage organisms and their relationships in ecological systems.

This chapter explains how partnership participants learn to manage nature, or in Latour's terms, how they "mobilize the world." Latour described the first loop of his conceptual model thus because the objects of scientific study are much more unruly than commonly appreciated, as the narrative opening this chapter makes clear. Scientific inquiry tries to bring order to the heterogeneity of nature, or at least to bring order to knowledge of its diversity. This chapter explains how partnership participants shifted their thinking from "technology transfer" to managing organisms and their relationships in dynamic farming systems. To "mobilize the world" requires knowledge of the behavior of organisms, how they are relating, and how farm-management decisions can create preferred social outcomes.

California's agroecological partnerships are particularly valuable as case studies because they allow comparative analysis. The 16 different commodities undertaking partnership have distinct agronomic, social, and marketing histories.[10] Within the 8 commodities hosting multiple partnerships, differing social relationships and epistemological assumptions have shaped different approaches to organizing agroecological strategies and practices.

This chapter defines agroecology by describing how partnerships draw from diverse knowledge sources to develop alternative practices. The balance of this chapter is organized primarily around the five chief partnership strategies listed in table 5.1, with commodity-specific examples.[11]

Table 5.1
The five chief agroecological strategies of California's agroecological partnerships ($N = 32$).

Parnerships facilitate agroecological learning	32
Parnerships facilitate agroecological pest and fertility monitoring	32
Parnerships develop agroecological pest-management techniques	31
Parnerships develop agroecological soil, fertility, and irrigation techniques	22
Parnerships facilitate agroecological integration of farming systems	22

These five are general to deploying agroecological initiatives in industrial agricultural systems. Any initiative to extend agroecology in America will have to deploy these five strategies to be successful, while helping growers shift their perception of risk.

Partnerships Facilitate Agroecological Learning

All partnerships try to facilitate agroecological learning, and all but a few of them have created a semi-formal process to enroll growers and pest-control advisors. The process of engaging growers—recruiting them and their participation in social learning—is absolutely critical to making progress toward agroecological goals. The variability in enrollment strategies reveals the assumptions partnership leaders hold about how growers behave and what motivates them. All 32 partnerships facilitated agroecological learning, and 30 devised structured social learning processes. Twenty-nine formally enrolled growers in the partnership and arranged for them to dedicate at least one orchard block/field for more than one year. Twenty-five required growers to devise comparison orchard blocks/fields.

The configuration of social learning and grower incentives varies by the biophysical nature of commodity production and the granting programs that fund partnerships. The understanding of enrollment varied by partnership, ranging from grower-led discovery learning to simply permitting pesticide trials to be conducted on their fields.[12] BIOS pioneers developed the formalized social learning process of grower-designated comparison blocks with a mix of practices selected from a menu by the grower. Partnerships create spaces for intensifying existing relationships by focusing social learning toward agreed-upon agroecological goals. Facilitating a shift in risk perception among growers has been critical to partnership success.

Three examples from the three pioneer partnerships illustrate the diverse meanings of enrollment. Doug Hemly enrolled other pear growers in the Randall Island Project informally, with only a verbal agreement, yet growers knew they were participating in experimental research on a field scale, and exposing their crop to some risk. Pheromone mating disruption requires the commitment of an entire contiguous block, and thus comparison blocks are generally inappropriate.

Leading Lodi winegrape growers persuaded a majority of their fellow growers in the region to self-tax to support an organization to conduct place-specific research and extension (plus the marketing of Lodi wines). These leading growers created an organization and became BIFS growers, hosting research and extension in their vineyards.

The BIOS partnerships fused populist ideals with agroecological practices, and tried to facilitate growers "taking control" of their orchard operations. They emphasized the voluntary nature of the partnership while providing educational materials to stoke growers' agroecological imaginations. These diverse forms of enrollment will have implications for how growers and other partners understand the scope of their activities.

The voluntary character of these partnerships is crucial to their appeal. BIOS leaders emphasized this non-prescriptive approach, and enrolled growers to set aside "BIOS blocks" in which growers themselves selected practices from a menu and compared results with a "grower standard" block of similar size. The grower was then able to decide which set of practices he judged better by his or her own criteria. This general approach has been copied—to varying degrees—by most California partnerships. Those partnerships focused only on pheromones did not use the comparison-block technique.

Almond BIOS set out the most ambitious expectations of enrolled growers, and its leaders sought to enroll a mix of growers who relied on agriculture for their livelihood but were interested in making changes in their farming system: some visible and respected, some smaller, and some with high chemical use.[13] BIOS's expectations were formalized by the grower signing an "Agreement of Understanding" that made the above tasks explicit, detailed the technical support growers can expect from the BIOS management team, and affirmed that farm-management guidelines are suggestions, not official recommendations. Once that agreement was signed, the management team conducted a farm visit to customize a management plan for the BIOS block.

In the pear and walnut partnerships, managing the codling moth is the most important objective. Conducting applied research on codling moth pheromones gave these partnerships a strong orientation toward eco-rational technology, but also on the benefits and necessity of area-wide cooperation. As the success of the Randall Island Project became known, pear growers in other regions asked for partnerships because they wanted to learn about mating disruption too. Walnut partnership leaders see the removal of organophosphates and the successful deployment of pheromones as critical to taking advantage of integrated farming systems strategies, and the search for pheromone-based pest-management strategies spurred the creation of four of the five walnut partnerships. This is second only to the number of winegrape partnerships. Partnerships deploying pheromone mating-disruption technologies have generally made greater efforts to enroll PCAs, recognizing the importance of their technical skills to partnership success.[14]

Social relations in the winegrape industry foster a high degree of grower dedication to partnership activities. California's winegrape industry has been consistently rewarded economically for cooperative actions, most clearly visible in American Viticultural Areas, or appellations.[15] Because winegrape growers are accustomed to sharing more information about their crop with wineries, they also appear to be more comfortable doing so with production data than growers in other commodities. Three later winegrape partnerships did not need to formally enroll growers and their vineyard blocks because previous partnerships had already established the viability of agroecological practices.

In sum, the enrollment of growers has had various operational meanings, ranging from a grower allowing research to be undertaken on his or her field to a grower creating opportunities for himself or herself and others to learn about alternative practices. Most partnerships have tried to enroll some growers' blocks and PCAs with them. The initial description of activities and expressed expectations about growers' participation frame the scope of agroecological experimentation and learning.

Partnerships Facilitate Agroecological Pest and Fertility Monitoring

Every partnership expected growers to learn more about monitoring their farming system. Leading architects of the agroecological partnership model believe that the lack of perception and understanding of agroecological organisms and relationships by growers (and their PCAs) is a major impediment to using agroecological strategies. All partnerships tried to help growers and their PCAs more fully perceive the dynamic behavior of agroecological organisms to help them better manage their farming system. All but one facilitated enhanced pest monitoring, and 16 facilitated enhanced crop, soil, and irrigation monitoring and analysis.

Specific examples of the monitoring practices are listed in table 5.2. Fifteen report developing new pest or fertility monitoring techniques. Partnership leaders determine the scope of alternatives they want growers to consider by what they want growers and their PCAs to monitor and count. Efforts to encourage increased or more sophisticated monitoring have been most successful when growers understand the process, have confidence in it, and perceive an opportunity to at least break even or save money.

Social learning about agroecological pest-management strategies proceeds through a graduated process of enhanced perception. The three main steps are these:

Recognize pests as an biological organism.
Perceive the insects in the context of a dynamic insect complex.
Understand the insect complex as one component of a dynamic farming system that can be managed.

The first step involves learning about the life cycle of the insects, especially its feeding and reproductive patterns, to determine when the insect is most vulnerable to a grower's intervention. Very few insects are pests, and managing them effectively requires an understanding of their specific

Table 5.2
Agroecological monitoring practices promoted by selected partnerships.

	Pest monitoring techniques	Crop, soil and water monitoring
Walnuts, pears, apples (9)	Pheromone-based pest (codling moth) traps; agroecological monitoring protocols; assess beneficial insects	Well water nutrient analysis; leaf tissue analysis
Grapes (wine, table, raisin) (8)	Agroecological monitoring protocols; assess beneficial insects; insect ID sheets; computer monitoring data software	Well water nutrient analysis; leaf tissue analysis; moisture analysis
Almonds, prunes and stone fruit (5)	Agroecological monitoring protocols; pheromone-based traps; assess beneficial insects; insect ID sheets; computer monitoring data software	Well water nutrient analysis; leaf tissue analysis; tree moisture analysis

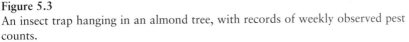

Figure 5.3
An insect trap hanging in an almond tree, with records of weekly observed pest counts.

life strategies, as the almond industry learned when it learned about the autecology of the NOW. Partnerships have developed insect-specific monitoring skills and insect identification guides for pest and beneficial insects. During the heyday of the chemical revolution, many growers and PCAs believed that "the only good bug is a dead bug," but most growers have since recognized the impracticality of this. Growers and PCAs in general recognize that "good bugs as well as bad bugs exist," but partnerships try to facilitate a more sophisticated understanding of the dynamic relationships of pests, their predators, and their agroecological condition.

The second step develops enhanced understanding of the agroecological relationships between various insects, especially predation and parasitism. These relationships are largely invisible to the untrained eye, so partnerships help growers and PCAs learn to see them by trapping, identifying and monitoring evidence of the insect pest and its natural enemies (figure 5.3). Partnership leaders hope this information can help

growers understand the potential of natural regulation of insect populations, especially once broad-spectrum organophosphates are removed
from the farming system. Partnership leaders hoped that by eliminating
organophosphates they could take advantage of new biological control
opportunities and reduce the expense of treating secondary pests. In several cases, however, when growers stopped using organophosphates the
partnership discovered that their broad spectrum action had controlled
insects that they had not realized were pests, or in some cases, they had
not even realized were in the farming system at all. Several partnerships
tried to calm the fears of growers by insisting that the insect complex in
a farming system would undergo a period of transition for several
seasons after the removal of organophosphates before biological control
agents could build up to effective numbers.

The third step presents the environmental complex for insects and their
natural enemies, and introduces to the grower possibilities for farm management that could shape the composition and behavior of insects.[16] By
managing crop plant health, non-crop vegetation (e.g., cover crops), and
on-farm habitat, growers can attract and maintain beneficial organisms
in the farming system and manage crop plants so that they are less
vulnerable to pest problems. Monitoring cultural practices such as sanitation is also a part of this stage. Partnerships ask growers to perceive
and understand the relationship between components of the farming system, and to recognize that how they manage their crops and non-crop
areas will shape the composition and behavior of the insect complex. For
example, by creating a healthier soil system, orchard trees can be less
stressed and less vulnerable to pest outbreaks, especially mites.

Partnerships try to help growers and PCAs advance through these
three steps by improving the quality of pest monitoring information,
whether done by growers, PCAs, or scouts employed by the partnership.
They educate growers and PCAs by demonstrating scouting skills at field
days, and by providing monitoring forms that convey more fully the
agroecological conditions within a block. (See figure 5.4.) Partnership
leaders believe that these activities will improve pest-management decisions in the short term, and enhance the understanding of growers and
PCAs of the dynamism of their farming system over the long term. Many
partnership leaders believe a fundamental obstacle in their efforts to
promote environmental stewardship in California agriculture is the dominant tendency to think of farming systems as passive and static.

Figure 5.4
Independent pest-control advisor Laura Breyer shows growers how to determine if a pest is causing economic damage at a Sonoma County Winegrape Growers Association field day.

Helping growers and PCAs to see not just a pest insect but an insect complex—including natural enemies—has been a goal of IPM for five decades. UC research conducted during the 1970s, the 1980s, and the 1990s about insect pests and their natural enemies—including improved traps, accurate monitoring protocols, and better population models—made new monitoring practices possible. Agroecological partnerships mark a new era in the application of ecological ideas in agriculture because a majority of them move beyond IPM to monitor and analyze the entire farming system and manage it in an ecologically optimal way.

Expert monitoring by PCAs or other skilled entomologists was and will continue to be critical to the success of the pheromone mating disruption. When pheromone-based products were first brought to market, few growers and PCAs had the agroecological skills to monitor and understand the population dynamics of the codling moth. Because mating-disruption technologies do not kill, the insect pest is usually still present in the orchard after an application, at a time when growers and PCAs were used to seeing them dead. Mating disruption requires confidence in monitoring population models and that breaking the life cycle of an insect can be as effective as killing a pest. Many of the new pesticides are not only specific to species but also specific to life stage (e.g., larva).

Most partnerships try to facilitate growers learning about the value of enhanced monitoring services. Almond BIOS used earlier UC research to create an agroecological orchard monitoring form—customized to almond farming—that included pest status (none/low/medium/high/treatable) plus beneficial insect presence, diseases, and tree health. In keeping with their populist ideals, the BIOS pioneers hoped growers would learn how to monitor their own orchard, but that goal was essentially abandoned when growers did not do it. Grower monitoring conflicts with the dominant division of pest-management labor in California agriculture. PCAs have specialized training, use it daily during the growing season, and regularly access experts who can help them solve unusual insect pest problems. BIOS staff had to content themselves with providing additional information to growers about how to become better consumers of PCA services, and hope that would encourage him or her to ask for more thorough information from their PCA. After its second year BIOS hired scouts (without a PCA license) or independent PCAs to gather field data. BIOS encouraged growers to consider using independent PCAs because they provided better information about the insects in the orchard, both pests and beneficial insects. Some growers switched to independent PCAs (and this did not endear BIOS to affiliated PCAs), but many growers were skeptical that they could save money by spending extra for a service they had heretofore received for free. BIOS promoted monitoring leaf tissue and irrigation water for nitrogen levels, and encouraged growers to adapt their fertility program in keeping with results from them. The BIFS partnerships working with annual and animal crops (rice, field crops, and dairy) developed simple, field-appropriate means to monitor nitrogen.

Persuading growers or PCAs to monitor systematically is the rock upon which many partnerships founder. Some partnerships created protocols and devices to help growers and PCAs monitor (pest-management protocols and treatments thresholds are covered in the next section). The

UC Berkeley entomologist and researcher Nick Mills has encountered numerous growers in orchards while monitoring the biological control agents he introduced, but he reports:

They're very interested in knowing more about what's going in their orchard and getting to recognize what some of the insects are that are there. . . . Somebody has to spend time going around and looking, and the growers love it. They love to see these reports of how few aphids they've got and how few moths they've got and all the rest of it. They love to see these things, but then do they want to do that monitoring afterwards? No. No. Does the PCA want to do that monitoring? No. Does the grower want to pay a PCA to do that type of monitoring? No. And it's unfortunate now, because there just is such a low return on agricultural production that everybody is looking for ways to cut expenditures rather than to [pay for monitoring]. So, it's very sad, really.

Monitoring information is not cheap. One chief reason agrochemical use in California continues to be high is that better knowledge of agroecological conditions costs growers money, money that many would prefer to spend treating a problem that might exist rather than discovering whether or not a problem in fact exists. In some cases, capturing data to treat a problem may cost more than the treatment. Partnerships try to help growers recognize the value of more efficient input use and its money saving potential. Putting agroecology into action requires more sophisticated information about the condition of ecological organisms in a specific farming system. Monitoring alone does not advance partnerships toward their goals, which depend on putting knowledge derived from monitoring into action.

Partnerships Develop Agroecological Pest-Management Practices

Approaches to addressing pesticide-related environmental problems fall into two main strategies: facilitating learning about strategies to integrate farming systems so as to reduce or eliminate pesticides, and demonstrating that new "softer" pesticides can effectively and economically be substituted for disruptive and hazardous broad-spectrum pesticides.[17] Generally, BIOS and BIFS partnerships have emphasized the former, intensive and more difficult approach, while PMA and pheromone mating-disruption-based partnerships generally tend toward the latter. Virtually all partnerships scrutinized the received wisdom about economic damage thresholds, and facilitated applied research demonstrating that many pesticide applications are or can be made unnecessary. Partnerships tried to communicate that conventional pesticides impact an entire insect complex, not just the pest in question, and that the value of reducing the population of that pest with an organophosphate had to be weighed against the likelihood that additional treatments would be needed as a result of the disruption caused. Changing pesticide use, in either quantity or quality, has been the chief goal of all but three.[18]

Most growers participating in partnerships share the expectation that organophosphates are "going away" at some point in the near future due to regulatory action, and participate in partnerships in part because they want to develop alternatives before they are "forced" on them. Partnerships funded or staffed by CAFF and SAREP have held out the ideal of pesticide reduction, while most grower participants take a more pragmatic approach of substituting pesticides as their management conditions dictate. Biocontrol and cultural pest-control techniques have played an auxiliary role. Specific examples of practices are listed in table 5.3.

Agroecological partnerships integrated social concern about pesticide use with the new dynamics within farming system insect complexes. Nine leaders cited organophosphate pesticide resistance as a contributing factor to the creation of the partnership. (Recall table 3.3.) This plus

Table 5.3
Chief techniques of insect pest management promoted by selected partnerships.

	Techniques to reduce and replace pesticides	Biocontrol and cultural techniques
Walnuts, pears, apples	Pheromone mating disruption; foster biocontrol by eliminating OPs from orchard; precise timing of pesticide applications; reduced rates of application	Orchard sanitation; beneficial insect releases; bird/bat boxes
Grapes (wine, table, raisin)	Educating growers and wineries about thresholds; softer pesticides; precise timing of pesticide applications; reduced rates of application	Leaf pulling; beneficial insect releases; cover crops to moderate vigor
Almonds, prunes, stone fruit	Develop specific economic thresholds; pesticide use decision rules; softer pesticides (Bt, insect growth regulators, pheromones, ant baits); precise timing of pesticide applications; reduced rates of application	Early harvest; orchard sanitation; beneficial insect releases; cover crops

the passage of the FQPA created a sense of crisis that stimulated growers and their organizations to recognize the value of organophosphate
alternatives, much as Earl Decker's network did in 1978. Reducing or
removing organophosphates from farming systems allows formerly sup-
pressed ecological organisms to express their potency, their life. Growers
and PCAs prescribed softer, pest species-specific pesticides, relied on
pheromones to reduce pest populations so that they relied on drastically
fewer organophosphate applications, used non-hazardous materials like
soaps and mineral oils, or discovered that much of the benefits they had
received from organophosphates mixed with oils could be garnered from
oils alone.

Newer, biologically derived, pest-specific pheromones and insecticides
have helped a great deal, and the passage of the FQPA accelerated their
development. UC IPM Advisor Walt Bentley observes:

. . . not everybody wants to hear this, but there have been pesticides that have
allowed growers to move away from what we would consider more harmful
materials. There are things like Spinosad [a pesticide product derived from a soil
micro organism that selectively kills plant eating insects]. We've found better uses
for Bt. We found [insect] growth regulators. These have all helped to be integrated into the program, so those growers that don't have the confidence of
growing without a spray can still rely on that, maybe a little bit of their old program and not have the destructive effects. And that was a key part of both the
almond and the stone fruit program. These programs came together when there
were materials that could fit the needs.

But many new, less hazardous pesticides cannot be used with the same
ease as the broad spectrum organophosphates; they are not interchange-
able, and they require support from a social learning network to be
successful.

Several general scenarios unfold as growers remove organophosphates
from their farming systems[19]:

New "softer" products provided some control of primary pests, but not
enough to justify their expense (many are quite costly the first few years
they are on the market).

Participants discovered that the organophosphate was controlling more
than the primary pest, and they had to develop strategies to cope with
the secondary pest(s), often turning to additional soft pesticides.

Some growers applied organophosphates every year (especially in dormant sprays) but discovered they were economically justified only every
2, 3, or 4 years.

Organophosphates were not as critical to pest management as commonly believed: this may be because the economic damage threshold was higher than thought or because biological control agents could provide some control when given the opportunity.

The removal of organophosphates results in one or more secondary pests becoming problematic for several years, but natural enemies or cultural techniques begin to provide some control after several seasons of transition.

A few partnerships report that some participating growers recognized the superiority of the alternative practices demonstrated in their test blocks and abandoned their "industry standard" replication before the partnership had run its course. This was not uncommon in commodities in which partnerships made substantial success in organophosphate reduction (almonds, pears). Collectively these scenarios demonstrate how little is known about the overall behavior of the insect complex in many crops, and justify the claim that pesticides are frequently over-prescribed.

Partnerships Develop Agroecological Soil, Fertility, and Irrigation Techniques

Twenty-two of the partnerships have promoted alternative soil, nutrient, and water practices. Nutrient pollution in California has not yet caught the attention of regulatory agencies or the public as it has in other regions, although nitrogen and persistent (pre-emergent) herbicides do contaminate groundwater in California. Most of the partnerships addressing soil, nutrient, and water management do so to improve resource-use efficiency, but also to take advantage of synergistic effects between farming system components. For example, scientists and growers found that dense cover crops hosted sufficient decomposing organisms to destroy almond mummies, and eliminate the need for and expense of flail mowing. Synergistic benefits from integrated farming systems are a further justification for the claim that agroecological partnerships are more than IPM efforts.

Cover crops are the most common tool partnerships use for helping growers improve their management of soil, nutrients, and water (table 5.4). Partnerships have promoted cover crops as the fundamental strategy for garnering a wide array of benefits for soil, nutrient, and water

Table 5.4
Soil, fertility, and irrigation management techniques promoted by selected partnerships.

	Soil and weed management	Fertility management	Water management
Walnuts, pears, apples	Cover crops; substitute contact for persistent herbicides	Cover crops; compost; nitrogen budgets; chipped prunings	
Grapes (wine, table, raisin)	Cover crops; substitute contact for persistent herbicides	Cover crops; compost and pomace application	Precision irrigation
Almonds, prunes, stone fruit	Cover crops; substitute contact for persistent herbicides	Cover crops; compost; leaf tissue sampling; nitrogen budgets; chipped prunings	Irrigation water analysis; leaf tissue moisture analysis; precision irrigation; buffer strips

management, but they must be planned, deployed, and managed carefully to achieve the desired benefits.[20] For years, organic growers have used cover crops as their primary strategy for soil nutrition. BIOS claimed that cover crops could enhance insect diversity and provide some pest control, but several UC scientists have tried to find direct evidence of this through research with little success.[21]

Managing cover crops is always about tradeoffs. They provide benefits, but being able to select the right variety for one's farming system and managing it appropriately are essential. Many growers describe having to re-think their approach to orchard floor management to successfully derive benefits from cover crops. Proponents of integrated farming systems insist that cover crops improve soil quality, enhance water infiltration, and provide nutrients, all of which improve crop health. Growers who have tried and abandoned cover crops describe them as providing marginal benefits but interfering with access to their crop, and demanding substantial management effort.

BIOS drew from Bob Bugg's ecological understanding of cover crops, but also turned to the applied expertise of Fred Thomas, a Chico-based consultant who participated in seven partnerships. Thomas has sold cover crop seeds and helped growers plan out how to use them for more than two decades, providing practical, hands-on advice for how to use them effectively. He is critical of most UC cover crop research because it either has not used the correct seed mix, or has not matched the seed mix to the soil needs, or is not replicated properly.[22] Thomas has a cover crop mix for every purpose. Legumes fix nitrogen. Tap-rooted bell beans and oats break up plowpan. Perennial grasses provide soil stability allowing tractor access during wet winters. A diverse mix of cover crop species can occupy the ecological niches of weeds. Specialized mustard species can reduce the populations of nematodes, microscopic soil-dwelling organisms that can torture orchard trees. For growers concerned about the potential of increased frost damage from cover crops he recommends species that can survive low mowing in the winter. For growers concerned about high irrigation costs he recommends mixes that go summer dormant. Thomas guardedly suggests that cover crops can aide in insect pest management by providing nectar, pollen, or alternate hosts for beneficial insects, but cautions that the generalist predators do not have the ability to control pests with explosive growth, such as prune aphids.

Thomas emphasizes that cover crops are not a panacea.[23] Cover crops can work wonders with farm soils, but they have to be tied to realistic biological goals and growers have to be able to adapt their approach to orchard floor management. No one cover crop mix can perform all functions. If growers want to improve winter access to their orchards, they should sow grasses, not tall growing leguminous plants. If they want an immediate fertility boost, they should plant legumes, although these grow tall enough that they can increase the risk of frost damage to trees in some regions. They also preclude equipment access during the spring, when chemical disease treatments can be crucial. Cover crops with the most nutrient value usually have to be purchased and planted each year. Permanent cover grasses allow equipment access in the winter, but require water during the summer. Permanent cover crops work well in orchards with flood irrigation, which tend to have affordable water, but they can be quite expensive for a grower trying to conserve water with drip irrigation. Some growers have balanced orchard access with other cover crop benefits by planting them in alternate middles.

To effectively incorporate cover crops into their operations, growers must be open to re-thinking some aspects of their farming and have the ability to match the variable performance of different cover crops to the

Table 5.5
Potential benefits and drawbacks of cover crops.

Potential benefits	Potential drawbacks
Improved soil structure	Increased water use
Improved water infiltration	Competition for moisture and nutrients
Improved tractor wheel traction	Increased frost hazard
Addition or conservation of nitrogen	Habitat for vertebrate pests
Addition of carbon and organic matter	Increased time needed to manage
Dust reduction to reduce mite outbreaks	Increased costs
Control of soil erosion and nutrient run-off	

site specific needs of their fields' soils. Growers seeking a single benefit previously provided by a single agrochemical product are disappointed. Growers who have developed the eye to see incremental, cumulative and multiple benefits over time have made cover crops a centerpiece of their soil and nutrient management.

Growers of winegrapes have found cover crops to be compatible with winegrape growing for several reasons, and the result is that 15–30 percent of them use cover crops.[24] Vineyards on hillsides have historically used cover crops to control erosion.[25] Grape vines have modest nitrogen needs—roughly 25 percent of the per acre requirements of rice or walnuts—which can be provided easily by leguminous cover crops. Researchers thoroughly documented the problems associated with excess nitrogen in vineyards: over-application of nitrogen to grape vines can result in excessive vigor, which stimulates canopy growth shading the fruit, creates a favorable microclimate for pests and disease, and decreases fruit quality. Cover crops, or at least resident vegetation, help this strategy to reduce vine vigor by providing competition. This is an example of the multiple, indirect benefits that cover crops provide.

The annual and animal crop partnerships, more than those in perennial crops, tended to put emphasis on soil, fertility, and irrigation techniques. The field-crop BIFS partnership developed new cultivation and soil fertility management techniques.[26] Pre-plant tillage accounts for more than one-fifth of total production costs, so the partnership analyzed various biological inputs and reduced tillage strategies. This partnership had to adapt conservation tillage principles to local conditions just as Kupers and the ACIRDS partnership did in Washington. The major thrust of the Rice BIFS was to research and demonstrate alternative soil and water management strategies to take advantage of rice straw nitrogen and reduce the need for herbicides. The fundamental goal of Dairy BIFS was to optimize the use of manures to fertilize their forage crops. Producers have to be able to see manure as a genuine fertilizer, and then be able to re-structure their manure pond plumbing so as to take advantage of it. The practices developed by dairy BIFS appeal greatly to producers who could manage manure without major capital investment.

Partnerships appeal to growers by providing opportunities to experiment with innovative soil, fertility and irrigation techniques, but these techniques often require the grower to re-think assumptions about how their farming system should function and appear. These cases illustrate the difficulties and opportunities cover crops present. To deploy them successfully requires growers to re-think some assumptions about industrial monocrop agriculture. This can best be accomplished by helping growers and PCAs think about their operation as an integrated farming system.

Partnerships Facilitate Integration of Farming Systems

The agroecological strategies of most partnerships distinguish their collective efforts from traditional IPM projects. Twenty-two partnerships facilitate agroecological learning, either by helping growers develop an integrated farming system management plan, or by promoting a systems approach with a manual or farming system assessment guide. Fifteen partnerships helped growers develop management plans, and twelve created partnership manuals. These strategies are designed to help growers take advantage of beneficial interactions between the components of their farming systems, such as optimal soil and nutrient use by crops to reduce insect pest susceptibility. These represent the most sophisticated use of agroecological principles in conventional California agriculture, and manifest the meaning of "integrated farming systems" here.

The BIOS pioneers developed the strategy of a management team farm visit to provide input for an integrated farming system management plan. Growers' appreciation for these plans appears to have varied widely within each of the ten partnerships that have used this strategy. BIOS created a template for integrated farming systems by presenting a complete menu of options but also a framework for learning how to integrate them effectively. It adapted practices and imaginaries from organic agriculture to make them practical for conventional growers. Most of the pest-control strategies promoted by almond BIOS were fully compatible with organic practices; growers who did not want to become organic cited disease and weed concerns. Since the agrochemical revolution, Cooperative Extension has promoted a package of techniques, sometimes known as Best Management Practices. In contrast, BIOS encouraged growers to perceive and think of their farm differently, as an ecological system, not as an assemblage of crop plants. The *BIOS for Almonds* manual presented an integrated farming systems approach based on four principles:

Feed the soil first.
Make a home for predators.
Keep your eyes peeled.
Work with nature.[27]

The first principle is taken from the organic farming movement. BIOS framed pest management as a farming activity fully integrated with other components of the farming system. Learning about pest management in the BIOS partnerships, at least according to the manual, was secondary to learning a new approach to thinking about the farming system. BIOS encouraged growers to approach their farming systems as an ecosystem as much as possible, and allow nature to work for them. The fourth principle refers more to an ideal mindset rather than any practices.[28] Thus, BIOS tried to emphasize a broad framework for learning about one's farming system and working creatively with its components. BIOS leaders would often have to resort to describing individual practices (e.g., cover crops) to explain how it differs from conventional farming, but they tried very hard to emphasize that it was more an alternative, integrated approach to farming than a collection of discrete alternative practices.

The *BIOS for Almonds* manual follows the farming system throughout the year, presenting knowledge in a way that is most meaningful to a grower's annual planning, following biological time. Leaders invested time and resources in these manuals because their partnerships presented an alternative approach to their respective farming systems. Partnerships proposing modest changes in one or two farming system components have no need to invest in creating a manual.

Not all partnerships need to take this systems approach to take advantage of agroecological strategies, however. Based on his experience in four pheromone mating-disruption partnerships, Pat Weddle does not ascribe to the systems approach used by BIOS. He believes that partnerships can make more important progress toward pollution prevention by focusing their efforts on one problematic practice. Weddle understands the systemic nature of chemical-intensive agriculture. Not a reductionist, he simply believes that change is more likely to occur when an intervention concentrates on a single obstacle to improvement.

Twelve partnerships developed manuals to facilitate a more comprehensive understanding by growers of these principles and how to apply them. The manuals listed in table 5.6 were written as companions to

Table 5.6
Manuals for integrating farming systems.

Crop	Manual
Almonds	BIOS for Almonds (CAFF 1995)
	A Seasonal Guide to Environmentally Responsible Pest Management in Almonds (Pickel et al. 2004)
Cotton	BASIC Cotton Manual (Gibbs et al. 2005)
Prunes	Integrated Prune Farming Practices Decision Guide (Olson et al. 2003)
Raisins	Sunmaid Best Management Practices Guide
Stone fruit	Stone fruit PMA Guide[a] (in preparation)
Strawberries	Organic Strawberry Production Manual (in preparation)
Winegrapes	Positive Points System (Central Coast Vineyard Team 2003)
	Lodi Winegrowers Workbook (Ohmart and Matthiesen 2000)
	Code of Sustainable Winegrowing Practices (California Association of Winegrape Growers and The Wine Institute 2003)
	Sonoma County Grape Growers Association IPM Fieldbook[a] (Sonoma County Grape Growers Association 2003)
	Integrated Pest Management Field Handbook for Napa County[a] (Napa Sustainable Winegrowing Group 1997)

a. These manuals address IPM issues only.

partnership activities, either to help participating growers or to present a comprehensive guide to the lessons learned by the partnership. Four focus only on pest management and are more accurately described as IPM manuals.[29]

The three winegrape partnerships using an integrated farming system approach understood their task somewhat differently than the BIOS manual described above. Their manuals are constructed less like a guide and more like an integrated business audit. These three are the most complete guidebooks for sustainability developed in California agriculture. They present objective criteria to growers to evaluate their own operation, and then link that back to group processes. They are a sophisticated and highly effective structure for facilitating social learning.

Winegrape growers and vineyard managers came together to improve their environmental practices in the Central Coast region, and in 1995 they formed the Central Coast Vineyard Team. They were motivated by environmental and land-use controversies, and were not directly influenced by the integrated farming system of BIOS. Their initial idea was to develop a best management practices guidebook for the region.[30] The group examined existing point systems scoring the adoption of practices, but made a significant conceptual switch: re-orient the design of the system to reward growers by "giving them points" for good practices instead of "penalizing" them for failing to do so, which has been the more common approach. This was an effective pedagogical shift. Although this is only one simple feature of their scoring system, team members report that its positive character makes it much more appealing. The creation of the Positive Points System was important as both a process and as a management tool, and both have improved the way the partnership participants understand their farming systems.[31]

In 1999, buoyed by the success of his BIFS program, Cliff Ohmart began the Lodi partnership's "Integrated Farming Program" in order to stimulate more systems thinking among growers. Its foundation is the *LWWC Winegrower's Workbook: A Self-Assessment of Integrated Farming Practices*. Ohmart and Matthiasson developed this workbook from several years of presenting this material in local growers meetings and at field days. They held "workbook workshops" to provide semi-structured social learning opportunities for growers and PCAs. In a "workbook workshop," growers are instructed to connect the name of a pest, a photo of the pest, and a photograph of the vine damage it causes, to connect photos of pests with overwintering and egg-laying sites, and to connect photos of pests with cultural practices and natural enemies that can control them without the use of pesticides. The workbook is a tool for systematic self-evaluation in every aspect of winegrape farming systems, integrating sustainable and quality winegrape growing practices. The workbook helps growers to measure the level of their adoption of integrated farming system practices, and to identify action plans to increase these.[32] It has laid the foundation for additional agroecological initiatives in Lodi and other California winegrape growing regions.

California's leading wine and winegrape industry organizations, the Wine Institute and the California Association of Winegrape Growers respectively, recognized the value of the Lodi partnership, the Central Coast partnership, and two other local partnerships. In 2001, these two statewide organizations partnered to develop a "Code of Sustainable Winegrowing Practices." Jeff Dlott guided its development, and he facilitated monthly meetings for a year and a half with 50 prominent vineyard and winery leaders. Its primary audience consists of vineyard managers and winemakers, but its sponsors are trying to appeal to all the people and institutions (neighbors, regulatory agencies, input suppliers, customers) with whom the industry interacts.

The Code adopted the four-point scoring system for viticultural practices in its entirety from the Lodi workbook. The Code takes a whole systems approach to winegrape growing, wine production and distribution, including social and environmental criteria. Grower outreach with the Code is through half-day workshops, in which vineyard and winery managers work through the book, learn about the criteria of sustainability promoted by the Code, and assess their own operations in light of this. It represents an integrated approach to "winegrowing," because workshop participants evaluate their vineyard and/or winemaking operations.

Agroecology and the Landscape of Risk

As growers increase their reliance on agroecological strategies they undertake new risks. Chemical-intensive agriculture entails risks, but generally produces a more consistent result on an annual basis than does more agroecological farming. The risk implications of agroecology in industrialized countries have not received the scholarly attention they deserve. Some agroecological partnerships do include discussions of risk, but very few treat it comprehensively. In the developing world, agroecological strategies help reduce the risk to farmers by mimicking natural ecosystems, by diversifying in-field cropping patterns or, incorporating animals into farming systems, to help farmers hedge their bets against uncertainties such as weather and pest pressures.[33] In industrial agriculture, the extra labor and reductions of efficiencies of scale would put such an approach at a competitive disadvantage.

Industrial monocultures integrate efficiently into economic markets, but expose growers to more agronomic risk because of their ecological instability. This production system generally increases the per acre production of one crop, but the resulting farming system is inherently brittle and unstable, depending on agrochemical inputs to remain productive. Specialty-crop farming and its distribution systems concentrate risks on growers because they must maintain a certain flow of marketable crops each year. The logic of monoculture thwarts serious consideration of farming systems that more fully mimic natural ecosystems. Consequently, growers here pursue input optimization for pollution prevention and cost savings.

Risk has economic and agroecological dimensions, but it also has individual and social implications. Many agroecological partnerships make reference to the threat of regulations to persuade growers that it is to their advantage to try new practices now, before the state cancels pesticide regulations or requires growers to monitor the impact of their practices on streams. Some partnerships encourage growers to perceive a more comprehensive set of risks associated with agrochemical use. Figure 5.5 represents the conventional wisdom that the greatest risk a grower faces is crop loss, and that expert knowledge, embodied in PCAs and Farm Advisors, is essential to managing it. Partnerships communicate an expanded array of risks to growers (figure 5.6). Crop loss and expert knowledge are still a part of the grower's consideration, but so are other risks posed by agrochemical use. Partnerships bundle these risks together and present them to growers so that they can perceive the full degree of risk posed by agrochemical use, and further enroll them in the goals of the partnerships.

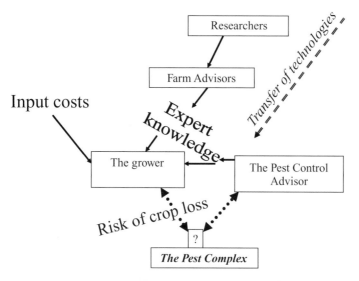

Figure 5.5
Conventional wisdom about the risks of pest-management decision making.

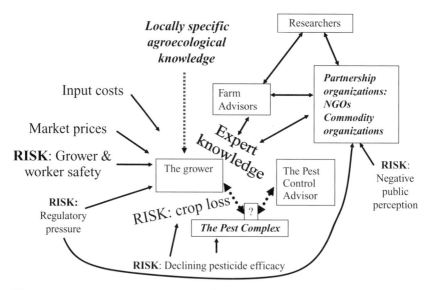

Figure 5.6
Partnerships promote a more comprehensive perception of risk.

A grower's risk position is defined by his or her farm-management style, economic resources, land tenure, and basic tolerance for uncertainty. Partnerships are not likely to change growers' fundamental stance toward risk, but may help them perceive more kinds of risk to their farming vocation. In addition to the commonly held perceptions of the risk that reduced agrochemical use poses to crops, growers may learn to perceive in a new way industrial agriculture's health, regulatory, and environmental risks. The multi-year transition process promoted by most partnerships is designed to help growers become comfortable with new risks. When decisions are made the economic incentives are rarely present for trying a new approach with the additional risk posed, and the potential for greater losses is always present. Growers and PCAs routinely make decisions throughout the season that bear million or multi-million-dollar risks.

All agroecological partnerships purposefully promote research, knowledge, learning, and knowledge exchange to manage risk. Growers without the knowledge and experience to take advantage of ecological relationships are in fact exposed to more economic risk through trying these practices without appropriate guidance. Many growers report that partnerships socialized risk by drawing on the expertise of others. Agroecological partnerships generally have not adequately addressed the risks involved in transitioning a farming system to utilize more agroecological strategies. Partnership leaders are aware of this as an issue, but have not articulated a framework for helping a grower analyze the way integrated farming systems reconfigure risk.[34] One of the reasons almond BIOS was so successful was that, in selecting from a menu of options, growers developed greater awareness of the risks they were taking, or could take, and where they could access the knowledge to manage those risks.

Technology-intensive monoculture heightens risk for extensionists, both public and private. Pest-control advisors prescribe pesticides when in doubt, and this is for them a practical risk-management strategy. Farm Advisors too are quite conscious of the risks growers take on their advice. When advising growers about alternatives to pesticides, they generally communicate the risks of pesticide reduction much more forcefully than any potential benefits of alternative strategies.

Conclusion

Agroecology in California takes advantage of the decades of investment in agroecologically informed research, exemplified by research into the biology of the navel orange worm, even though ecological strategies exist in tension within the dominant agricultural production system. Partnerships showcase agroecological growers and scientists and their ideas. They provide a structured social learning process that facilitates learning about the practices and the agroecological principles that underlie them. The five chief partnership activities define the domain of integrated farming systems in California.

According to Latour's model of circulating scientific knowledge, putting agroecology into action requires humans (growers, scientists, organizations, and agencies) to mobilize their efforts around the behavior of organisms in commodity specific farming systems. Partnerships are configured to help these participants learn together about the ecology of crops, insects, nutrients, water, soil, and technologies.

The first of the five chief partnership strategies is to enroll growers in an alternative agriculture project. The more successful partnerships adopt a systems approach and persuade growers to consider a wide range of changes.

The second strategy helps growers to value more precise readings of the agroecological conditions in their farming systems through monitoring. Growers generally do not monitor, but do appreciate quality monitoring reports, although they may not fully understand how these reports can translate into cost savings for them.

The third strategy is to deploy agroecological approaches to pest management, building on California's IPM tradition. All partnerships promote alternative pest-management techniques that reduce environmental impacts, and most present a menu of options, allowing growers to do their own learning about them. These alternative practices fall into two main categories: facilitating education about strategies to integrate farming systems so as to reduce or eliminate pesticides, and demonstrating that new "softer" pesticides can effectively and economically be substituted for disruptive and hazardous broad-spectrum pesticides.

The fourth strategy helps growers to think about opportunities for managing their soils using agroecological principles. A majority of partnerships promoted strategies to help growers perceive the relationship between ecological processes in the soil and crop plant health using cover crops.

The fifth strategy takes advantage of synergistic benefits by integrating components of farming systems. Partnerships help growers develop integrated farming system plans and provide educational opportunities through workbooks and workshops. These represent the most sophisticated use of agroecological principles in conventional California agriculture. Winegrape partnerships have used these strategies more effectively than those of other commodities.

To put these strategies into action, growers and their consultants have to be able to recognize how they provide benefits to their farming systems, and how their risks can be managed. Social networks, as described in the next chapter, are essential to managing those risks with increased knowledge, and to helping convert agroecological theory to action.

6

Agroecological Networks in Action

Sex and Sweetness

They had us humans right where they wanted us. How often in evolutionary history does one organism convince another to flood their entire habitat with the smell of sex? To develop a scientific sub-discipline to investigate it? And to manufacture sex pheromones on an industrial scale? And on top of that, to convince others to synthesize and release the sweet scent of pear fruit volatiles? To experience the joy of codling moth sex pheromones and pear kairomones, the Sacramento River pear district was the place to be!

Scientists assumed the perspective of males. Through evolutionary processes, the female codling moth developed the ability to use a species-specific chemical to attract the male, which in turn had developed specialized receptor cells in their hair-like antennae to transmit micro-electric impulses through the nervous system to the insect's brain and orient it toward the female for sex. Lepidopteran insects so fascinated scientists that in the 1950s they had developed an electroantennogram to measure the electrical current passing along the male antenna from the single sensory receptor, discovering the species-specific stimulus of these chemicals. So powerful are codling moth sex pheromones that they have persuaded us humans to stop trying to kill the insect and instead, deploy techniques to over-stimulate them.

Early laboratory research into insect behavior revealed the degree to which they as a taxonomic class relied on chemical stimuli for communication, feeding and reproduction. Scientists gave them a name: semiochemicals. Pheromones are semiochemicals for intra-species communication, or at least that is how some species of insects use them. In

1962, only a few years after their discovery, scientists started using codling moth pheromones to investigate sex in the field. George Post, who was then a Farm Advisor, recalls his first use of a monitoring trap with live female moths in a screened cage, which helped him to follow codling moth adult abundance over time. He discovered that the industry-standard spray after pear petal fall was entirely pointless because no young codling moth larvae were present in the orchard at that time. This led him to believe that to truly serve growers' needs, technology requires the support of expert knowledge. He quit his position with Cooperative Extension and became the first independent consultant in the Sacramento Valley, before the state even began issuing PCA licenses.

When DDT was introduced, the University of California's recommendations for spraying were based on the phenology of crops, not insect pests, and growers would apply them on a calendar basis, generally every other month during the growing season. This would kill all the insects in the orchard, and leave a residue on the fruit that would kill codling moth larvae after they emerge from eggs but before they are able to enter the fruit. The codling moth is thought to have evolved with apples in the mountains of what is now Kazakhstan, and migrated with that fruit to what is now Europe, where it found the pear and walnut also to its liking.

The codling moth co-evolved with the apple fruit, but readily damages pears and walnuts. The moths over-winter as full-grown larvae in tree bark, old branches, or in the soil at the base of the tree. Once temperatures reach 55°F, they pupate, emerge, and mate. A few weeks later, females lay up to 70 pinhead-size eggs in a scattershot on fruit, leaves, or calyx (clusters of fruit stems). Larvae emerge from eggs, and within 24 hours find and burrow into the developing fruit. From that point on, this generation is home free—pesticides cannot reach inside the fruit. When they emerge as adults, they begin again the cycle of mating, laying eggs, and burrowing into fruit. The development of both fruit crop and insect pest are regulated by environmental heat, and most seasons a grower can apply a pesticide based on the calendar date, but not consistently enough to gamble one's crop on it.

The same US Department of Agriculture IPM initiative that paid for the Ballico/Formosa Project in almonds funded Extension Specialist Clancy Davis at UC Berkeley from 1973 to 1976 to develop an IPM approach to pears, based on recent research that had been done on codling moth life stage models and plant pathogen predictive models. Aphids, scales, psylla, and mites join codling moth to attack pears. After DDT was banned, the standard pest-management regime was a toxic jumble of fourteen different organochlorine, organophosphate, and carbamate pesticides that wreaked havoc on natural enemies. Davis set out to rationalize this. He developed a codling moth control strategy using azinphosmethyl in low dosages timed precisely to be on the tree when the larvae hatched and migrated to fruit. He drew on earlier research that demonstrated codling moth could be controlled with half the label rate that would allow the survival of natural enemies for other pests. For Davis's plan to work, growers were going to need to know a lot more about the status of pests in their orchards, e.g., when they were biologically vulnerable to pesticides. The first step was to help "field men" understand how to use monitoring traps to understand the development of the codling moth.

Davis was able to use some of his USDA funding to hire pest-control experts and train them with his method, working with George Post and Pat Weddle, among others. Weddle was finishing his graduate work in entomology, and began working with Doug Hemly's father, who had himself collaborated with Cooperative Extension researchers for many years. Davis structured the program so that growers would be reimbursed for the expense of employing pest monitoring experts: 75 percent the first year, 50 percent the second year, and 25 percent the third year, inculcating them with the value of monitoring and scientific entomological knowledge. This program was so successful that it played a role in the creation of the PCA licensing system. As this project was drawing to a close, Davis codified this knowledge into the first crop-specific IPM manual in California.[1]

The codling moth is a major pest of apples, pears, and walnuts, so its food and sex practices have received an extraordinary amount of scientific attention. During the 1980s, scientists researched the precise chemistry of the sex pheromone and the behavioral response of males. Once evening temperatures reach 62°F in the spring, females release a pheromone plume that can be detected up to one-fourth of a mile. The males zigzag the boundary of the plume until they are within a few millimeters and make visual contact with the female, when they initiate mating behaviors. Moth pheromones are complex blends of several volatile oils. Male moths demonstrate a partial response to any moth blend, but require the precise, species-specific blend to stimulate the full sequence of mating behaviors. The first commercial application of manufactured codling moth pheromones was in monitoring traps, replacing the caged females. Scientists and pest-control advisors had to determine how much pheromone to put in each trap and how long it would last in orchards. They had to determine where to best place the product in the orchard. Most PCAs hang pheromone traps at the height of a pickup truck's window, but they should be placed high in the tree.

The codling moth populations in the Sacramento River pear district and the apple growing regions of Washington State began to develop resistance to azinphosmethyl, about the time a biological pesticide company released the first synthetic pheromone dispenser. In the laboratory, Isomate twist ties worked well on individual males, but because they are based on modifying collective behavior rather than killing the insect, scientists could not draw valid inference from their performance in test plots. The pesticide-resistance crisis provided an opportunity for scientists to study codling moth pheromone mating disruption on a field scale. Scientists and PCAs were able to draw some from their understanding of how pheromones stimulated insect behavior, but they had much to learn about how to manage an orchard flooded with pheromones. The volume and concentration of pheromones in Doug Hemly's orchard was without precedent in evolutionary history, and it was studied intensively.

Scientists developed several products to confuse the male codling moth. They found ways to put sex pheromones in plastic tubes, aerosol cartridges, wax flakes, chopped fibers, and microencapsulated oils. The twist ties worked best when hung in the upper third of the canopy. Harry Shorey of UC Riverside modified a bathroom deodorizer to create an aerosol device to release pheromone clouds (figure 6.1). These "puffers" were more expensive, but could be placed more widely in an orchard. They are less effective when orchards have high edge-to-area ratios. Puffer failures were quite costly, and later versions were controlled by computer chips to verify operation and release pheromones only during the twilight hours when the females call their mates. Pear trees are now pruned to a height of about 15 feet to reduce workplace liability for farm laborers on ladders. This results in a more compact orchard volume, and

Figure 6.1
Walnut BIFS grower Chris Locke explains to a neighbor how an aerosol pheromone dispenser disrupts the mating of codling moths in his orchard.

makes mating disruption easier. Walnut trees are two or three times this height, and have the greatest volume of any orchard crop in the state. Scientists developed wax emulsions to squirt high into the canopies of walnut trees, but they are most enthusiastic about developing microencapsulated pheromones that growers can spray with a conventional spray rig.

Early failures in preventing mating were determined to be from mated females invading from neighboring orchards, giving added weight to cooperative pest-control efforts. With additional research, scientists realized they were not fully preventing mating, but reducing it and delaying a significant portion of it. When the second-generation eggs hatch over an extended period of time, the resulting adults find fewer mates because they are spread out over time, which can result in a general decline in population numbers. Orchards with high populations of codling moth cannot be managed with mating-disruption technologies alone. The transition to successful mating disruption requires a conventional insecticide regime for a whole season, plus a vigorous sanitation program, including the removal of bins, brush, unpicked fruit, wood piles, general debris, and abandoned apple, pear, apricot, quince, crabapple, hawthorn, and walnut trees, to prevent re-infestation. In a few cases, the pest populations were so high that it merited an organophosphate application *after* the fruit had been picked to reduce populations to the point they could be managed with pheromones alone the following spring. Historical non-pesticide tactics included the use of corrugated cardboard (size Flute A, 18-inch rolls) collars around the base of trees to capture pupae in the winter, which are removed in the winter and burned. Some organic growers have used this effective, albeit labor-intensive, supplement to sanitation, and a few conventional growers are considering it again.

Insecticides leave undeniable evidence: dead insects. Initially scientists and PCAs had to monitor the effects of mating disruption on pest populations by inspecting fruit for damage, but very quickly they developed a "supercharged" monitoring lure with ten times the pheromones. This trap emitted chemicals that would "punch through" the background level of pheromones. It worked well most of the time, but that is not good enough for production agriculture, so researchers developed a kairomone lure that mimics the volatile odor of ripe Bartlett pears. PCAs now typically use these in tandem.

In recent years, scientists have investigated various explanations of the precise mating-disruption mechanisms. The false plume theory holds that the males follow the odor boundary of the manufactured pheromones to a dead end. The camouflage theory suggests that the manufactured products drown out the female pheromones. Some had proposed the sensory overload theory: the volume of pheromone in an orchard is so great as to overload the senses of the males and disorient them completely. The success of the supercharged lure confirmed that insect antennae adapt and filter out levels of overwhelming levels of pheromones. Finally, the distortion theory holds that the pheromone product disorients the navigational capacity of the insect. As of this writing, scientists agree that disruption is complex and observe that better control is achieved when mating-disruption technologies take advantage of several mechanisms.

The Randall Island Project was one of the first efforts to deploy synthetic coddling moth pheromones at a field scale, but Welter collaborated with researchers in the pear and apple growing regions of Washington, Oregon, and California. In 1994 and under the leadership of the Clinton administration, the USDA created a funding program to support area-wide IPM initiatives. Based on the promising results of the Randall Island Project and other preliminary efforts, the USDA selected the codling moth area-wide program (CAMP) for 5 years of funding. The project began on 2,630 acres, but with grower demand it expanded to 20,750 acres by the end, with 17 sites of at least 160 acres in four western states. CAMP provided technical support and shared costs of the pheromones. In 2001, about half of the pome fruit in the Pacific states was under mating disruption. Organophosphate applications during the CAMP program dropped by 80 percent. Thanks in large part to CAMP, pheromone mating disruption is now practiced on 100,000 acres in Washington.[2] Funding at the right time, for the right technologies—and supported by social learning processes—can make astounding environmental progress in agriculture.

Ecorational technologies fail without a social learning network to support them. Pheromone mating disruption is knowledge-intensive because the grower—or more likely, PCA—has to be able to monitor pest populations and the impact of pheromone technologies. Those monitoring mating disruption must be able to do the following:

identify the biofix (initial date of mating and accumulating degree-days) and anticipate when the first generation of adults will mate so as to disrupt them

interpret pest population dynamics across time and orchard space from counting trapped insects at single points and returning as frequently as once a day during critical summer periods

monitor the impact of disrupting the first generation on the second generation when they hatch, roughly 1,000 degree-days later (2 or 3 months), and determine if a supplemental pesticide is needed to kill larvae of the third generation

ensure that pheromones have disrupted the vulnerable spaces in the orchard by inspecting for fruit damage along edges and high in trees

verify the consistent performance of monitoring traps by visual inspection of the orchard to ensure they are not reporting "false negatives"

repeat monitoring the impact of any pre-harvest pesticide applications and predict the potential need for a pesticide application post-harvest or early spring to reduce codling moth population the following March, 7 months later.

Mating disruption will fail if growers or PCAs do not have a graduate-level understanding of insect population behavior, and the potential and limitations of monitoring data. Presence/absence monitoring will not work. Applying the materials and assuming they work risks the crop. Failing to verify the performance of the monitoring traps has resulted in the loss of tens of thousands of dollars of fruit. Monitoring codling moth mating requires an attentive scientific support network.

Several growers who have used pheromones methodically over multiple years report their codling moth populations have declined to the point where they are rarely detected. The extra expense of the codling moth pheromone products is offset by the biological control of other pests which formerly had needed additional pesticides because the organophosphate used on codling moth disrupted natural enemies. The Pear Pest Management Alliance funded additional work on biocontrol opportunities that presented themselves in an organophosphate-free orchard. Hemly and others hoped that pheromone products would eliminate the need for organophosphates, but he still finds that every few years they have to apply one at the first hatch to keep the populations low enough to allow the mating disruption to work.

The geography of California pear production played a facilitating role in the success of the Randall Island Project, and subsequent pear partnerships. Pears thrive in the moist soils of the Sacramento River Delta, and the narrow strip of orchard paralleling the river is the largest all-Bartlett growing region in the world. It currently produces about half of the state's pears. This is the southernmost and hottest pear district in the United States, and growers here had to manage an extra generation of codling moth most years. Thus, pear growers historically used azinphosmethyl in large quantities in this district, attracting Clancy Davis's attention. The geography of the pear commodity chain favored partnerships formation to increase the adoption of pheromone mating disruption (figure 6.2). California's other pear districts are concentrated along alluvial strips in Mendocino and Lake Counties, although a few Sierra Foothills orchards remain in production.

Other pear districts began clamoring for help funding pheromones about the time Jean-Marie Peltier moved to the Department of Pesticide Regulation and started the PMA program, which funded the California Pear Advisory Board in its first round. In 1995, the Environmental

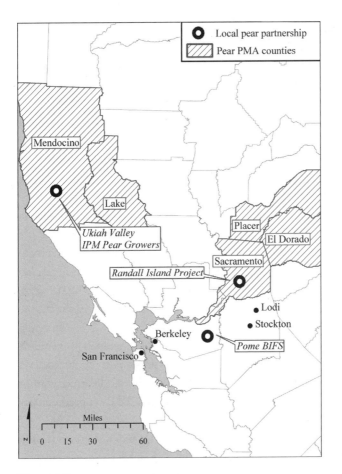

Figure 6.2
Map of California pear partnerships.

Working Group had tested some Gerber baby food and discovered small levels of pesticide residues. Sales dropped $50 million overnight. Gerber learned that it too has an interest in reducing pesticides use among suppliers, and now subsidizes pheromones. Pencap, the other organophosphate pear growers had used along with azinphosmethyl, was cancelled shortly after the passage of the Food Quality Protection Act in 1996. The California Department of Pesticide Regulation also substantially extended the re-entry interval for this material. The loss of these agrochemicals spurred additional interest in the use of pheromones to disrupt mating.

Other partnerships have emerged to help growers develop the skills to use pheromones to disrupt mating. In Michigan, the Center for Agricultural Partnerships helped organize and find funding for codling moth mating disruption in apples. David Epstein, the project's manager, recognized the importance of additional knowledge support for growers, but learned from other partnerships that when free monitoring is provided, some growers do not learn about the limitations of non-lethal pest-control strategies. His partnership invested more resources in orienting growers to this new strategy and less in free monitoring as an incentive for adoption. He reports: "in my opinion, any time you are attempting to . . . change social behaviors, you have to do intensive network building." This partnership had a substantial effect on the Michigan apple industry, expanding pheromone mating disruption from 850 acres in 1999 to 8,300 acres in 2001. The CAP partnership is no longer active, but most of the grower-led networks to coordinate area-wide mating disruption still are. Gerber is still involved in subsidizing the technology. Prices for apples have been terrible, and Michigan apple acreage has dropped by 25 percent, but the tenfold increase in pesticide resistance has made this pest-management strategy relatively attractive. CAP organized a mating-disruption partnership for California walnut growers, but the volume of orchard space that must be disrupted continues to be an economic obstacle.

Pheromone mating disruption is an eco-rational technology requiring networks of expert scientific support without precedent in agriculture. This technology will fail economically without sophisticated understanding of insect ecology and its dynamic behavior across time and space. It contrasts vividly with the simplicity of spraying a pesticide. These insects (at least the males) may die a happy death, but without a constant support of expert knowledge, growers expose themselves to serious loss.

Knowledge Networks

The human actors described in chapter 4 cannot successfully deploy the agroecological strategies detailed in chapter 5 without the kind of dynamic knowledge exchange that requires networks. Agroecological partnerships constitute networks which facilitate the generation of new forms of knowledge to better manage organisms and integrate their ecological relationships in farming systems. This chapter describes how these actors are configured into networks, and how partnerships in turn shape the roles and relationships of those who participate in them.

Much conventional social-science research has described agricultural extension as a simple, linear process: scientists develop expert knowledge and technologies, and then convey these fixed objects to uniformly receptive growers. This indeed may be appropriate for explaining "Transfer of Technology" activities, such as introducing pesticides, but it does not reflect the complex social learning processes of extending agroecology.[3] Single factor explanations of grower or scientist behavior must give way to more sophisticated understanding of their social networks, in much the same way that more systematic, ecological frameworks replace scientific research investigating single organisms. To interpret the extension activities of agroecological partnerships, we have to consider the role of all participating actors and the dynamics of their relationships.

Network analysis emphasizes the importance of the dynamic and shifting relationships between people, nature, and technological objects. A network approach shifts the focus from scale to connectivity, and conveys the multiple dimensions of relationships between people, institutions, nature, and technologies that ultimately shape change on the agricultural landscape. Bruno Latour, Michel Callon, and John Law developed actor-network theory as an STS methodology precisely because existing scholarship did not adequately explain the network dynamics of human and non-human relationships, especially when they involve scientific learning, scientific controversies, or environmental resource conflicts.[4]

Neva Hassanein's work on Wisconsin's intentional rotational grazing was the first to specifically address the role of networks in extending alternative agriculture.[5] She analyzed thirty networks of local farmers and graziers, and using sociological methodologies, she interpreted these as a social movement. Wisconsin's networks have a strong grassroots, populist flavor, even as they collaborate with CIAS, and many of them embrace alternative agricultural science (including organic) and markets simultaneously (table 6.1). In contrast, California's agroecological partnerships, with a few exceptions, cannot be described as a grassroots

Table 6.1
Wisconsin and California alternative agriculture networks.

	Wisconsin sustainable agriculture networks	California agroecological partnerships
Focus	Local knowledge about farming systems, sometimes organic	Agroecological knowledge; commodity specific
Origins	Draw on tradition of agricultural populism, and increasing social power of farmers, fostering rural community	Rooted in alternative pest management strategies (biocontrol and IPM) to develop alternative agriculture
Size, scale	Specific county, or multi-county focus; emphasis on local knowledge	May be one county, regional, or statewide
Purpose	Help farmers develop the skills to farm the way they want to, including the development of new markets	Fostering better scientific land managers; strong focus on organizing applied science work
Funding	Minimal, little needed for grassroots activities	Needed for professional science staff and activities
Role of pollution	Environmental problems are one motivating factor	Environmental regulatory problems are key to understanding them, largely due to toxicity of OPs
Staff	Volunteers, often farmers	Paid, professional, often scientists
Relationship with science institutions	Defending their own, grassroots autonomy; scientists contribute on farmers' terms	Strong focus on enrolling scientists and extensionists in the agro-environmental problem solving
Relationship with grower organization	Grassroots and local leadership; resisting the power of national commodity organizations	Strong ties to, and in many cases, sponsorship by state or national commodity-specific marketing orders
Role of environmental regulatory agencies	Very little if any	A major player behind the scenes; have provided millions of dollars to achieve agency goals with less controversy
Connected to markets	Many working to find direct marketing alternatives to the mass commodity markets	Conventionally, but emerging interest in specialized labels and eco-labels
State elected officials	Somewhat responsive to the expressed needs of these rural networks	Little interest on the part of a legislature dominated by urban interests

or a social-movement phenomenon. They are better understood as socio-technical networks, meaning diverse stakeholders working collaboratively toward common social and scientific goals. Socio-technical networks are more likely to be successful among large, or capital-intensive, or specialty, or Sunbelt agriculture than the populism of Wisconsin's grassroots networks.

California partnerships have not created new relationships so much as intensified existing ones. By tapping external resources, partnership leaders are able to offer incentives for collaboration to a wide array of participants, specifically for the generation of new applied knowledge or the hope of additional financial and professional gain. Participants described partnerships as strengthening their relationships with each other through knowledge exchange, or in terms of Latour's model, circulating knowledge through hybrid socio-technical networks.

To secure the participation of varied participants, partnership leaders have to configure incentives for exchanging knowledge within a network. The partnerships that agricultural leaders have deemed "good" or successful" have been facilitated by leaders who understand this. The balance of this chapter describes the network building strategies of several partnerships. It presents four pairs of partnership comparisons to contrast differing network strategies: differing motivations of NGOs and commodity organizations; the relative importance of leadership by growers or extensionists; informal grower associations and extensionists; the challenges and opportunities of growing a premium crop in regions highly valued for real estate development; and the relative advantages of local and statewide partnerships. These comparisons will demonstrate the range of network building strategies and the trade-offs for choosing one over the other, using sociograms to illustrate the relative strength of knowledge sharing relationships within these networks.[6]

Each sociogram has one social group or organization circled, representing the primary, lead actor in the partnership. The primary relationship supported by the partnership is identified with a triple line. Other relationships critical to partnership success are represented by a double line. Single lines indicate relationships strengthened by partnership activities, and broken lines show existing relationships acknowledged by but marginally important to partnerships.

Convergent and Divergent Motivations of NGOs and Commodity Organizations

The configurations of the aforementioned networks exhibit remarkable variation, reflecting the priorities of leaders who set them up and participation of various actors. Partnership leaders evince their strategies for inducing change by how they organize these networks and their relationships. The creation of BIOS marked the entrance of new institutional actors into extension activities in California agriculture. The Community Alliance with Family Farmers integrated environmental concerns and populist values with practical and economic information about agricultural alternatives. This "intrusion" was highly controversial among Farm Advisors: some attacked BIOS as illegitimate science, while a few supported it quietly. CAFF sought to enroll other institutions in this project, and was most successful in helping the Almond Board of California (ABC) to adapt some elements of BIOS.

For all its farmer-centric discourse, CAFF effectively was able to weave the BIOS network around itself, effectively capturing influence from other actors in agriculture, specifically Farm Advisors and affiliated PCAs. As figure 6.3 shows, CAFF's relationship with growers became the central dyad in this network, supported by the power of non-crop ecological actors and the scientific expertise. Through developing close ties with growers, CAFF addressed some of growers' economic and environmental problems. Those Farm Advisors and affiliated PCAs not

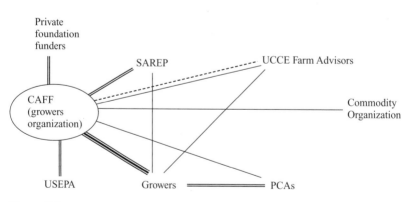

Figure 6.3
The almond BIOS network.

participating in BIOS correctly understood themselves to lose influence among BIOS growers: jealous Farm Advisors found themselves having to compete with a new interloper, even as CAFF worked closely with Farm Advisor Lonnie Hendricks. This is indicated by the parallel solid line and dashed line representing participating and non-participating Farm Advisors. CAFF drew new funders in to agricultural extension: the USEPA and private foundations.

CAFF deliberately reached out directly to the Almond Board of California when BIOS was not even 2 years old. Just as CAFF promoted the practical advantages of BIOS to growers, it appealed to the Almond Board's political pragmatism. Their strategy to engage a commodity organization was novel because it sought to influence a strong and influential organization of conventional growers. CAFF first built a relationship with the ABC's Production Research Committee by persuading it to co-sponsor a BIOS manual and make this approach available to a larger audience of growers.[7]

The Almond Board of California was not important to BIOS, but CAFF played a critical role in helping its staff recognize the value of participating in partnerships. In 1998, as BIOS was coming to a close, the DPR announced its new PMA grants program. CAFF's Marcia Gibbs in particular saw the opportunity to continue the good work begun by BIOS through further leadership by the ABC, and she encouraged the ABC to apply. Chris Heintz, the ABC's director of production research and environmental affairs, initially expressed reluctance because this grant would require the ABC to undertake new roles. Heintz hesitated because this would represent a qualitative shift in the ABC's relationship with growers, and she was unsure whether her board would approve this. Ultimately, and with help from ABC consultant Mark Looker, Gibbs helped Heintz recognize how the ABC could serve their growers helping them voluntarily adopt practices to avoid regulatory conflict. With the advice of UC IPM director Frank Zalom, they applied for a PMA grant.

Farm Advisors initially had highly ambivalent feelings about the PMA because they felt that CAFF had taken credit for (and channeled off grant money from) work that the University of California Cooperative Extension had done. Heintz had heard regulatory agencies repeatedly praise CAFF for its work with BIOS, and realized that their application would be much stronger with them as partners, even though it would irritate Farm Advisors. Once the grant was approved, Heintz and Looker convened a meeting of these Farm Advisors, assured them that they would not allow CAFF to claim credit for the UCCE's work, and explained to them the benefits this partnership would bring to the almond industry. They arranged the following division of labor:

Heintz was the principal investigator and Looker would be the project manager, running its daily operation.

Three Farm Advisors would recruit growers, set up the field demonstration sites, supervise the field scouts, and act as host for the field days.

University of California IPM director Zalom (along with IPM Farm Advisors Bentley and Pickel) would serve as science experts to the Almond Board and to the Farm Advisors.

The Almond Board and the Community Alliance with Family Farmers would jointly conduct outreach and education. CAFF would organize the field days and would create newsletters, which the ABC would publish and mail.

The Almond Hullers and Processors Association would assemble a master mailing list of all almond growers in the state for the newsletter.

Looker worked with the trade and local press to publicize the results of the field trials.

The ABC provided bookkeeping and cash flow (some smaller commodity organizations struggled to provide this support for their partnerships).

Staffers from the Department of Pesticide Regulation would liaise with the partnership management team.

Unlike BIOS, the almond PMA was oriented toward the UCCE and the ABC. Heintz has described the almond PMA as "a network of the leaders in the almond industry," which she specified as consisting of the Almond Hullers and Processors Association, the UC Cooperative Extension, the UC IPM Program, and the Department of Pesticide Regulation. She did not identify growers as partners. Once PMA funding ended, the Almond Board of California began using its own money to fund additional outreach and to create its own guide, *A Seasonal Guide to Environmentally Responsible Pest Management in Almonds.*[8] Heintz described the PMA as creating a structure for the ABC to demonstrate 30 years of research, accompanied by economic analysis. The PMA grant provided resources and gave the ABC the focus to do this. The ABC has internalized the goals of the PMA grant more than any

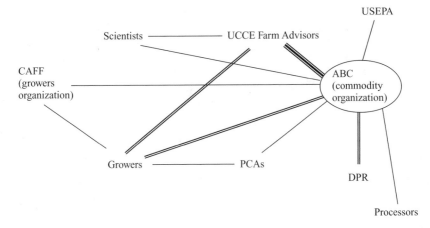

Figure 6.4
The almond PMA network.

other commodity board. This required the ABC to assume new roles with new responsibilities.

According to Looker, the Farm Advisors were crucial to the success of the almond PMA:

To me, they are the heart of the PMA and what makes it happen. They are on the ground, they know the growers, they know their local area, and without them, there is no PMA. The Almond Board couldn't go in on its own and make this thing happen. . . . They have a relationship already, they are seen as objective and impartial. All the UC guys I know are very well respected by their grower community. That relationship, that bond of trust is already there.

At the same time, the three Farm Advisors found the PMA useful because, according to Looker, it enabled them to mail materials to all 5,000 almond growers in the state, to do applied problem solving with growers, to receive positive press coverage of that, to contact growers they would not otherwise reach, and to stimulate positive feedback from the industry to their UC supervisors.

The almond PMA has appealed to Farm Advisors in part because the ABC stepped in to assume some of the activities conducted by UCCE prior to its budget cuts. The almond PMA made Farm Advisors' work more efficient while destabilizing neither their position as experts, nor their relationship with growers. At the same time, the ABC was able to develop stronger ties with its growers than ever before. The ABC worked with the Almond Hullers and Processors Association to persuade indi-

Figure 6.5
Almond growers at a Pest Management Alliance field day in Escalon, California.

vidual processors to turn over names and addresses of "their" almond growers. This was a historic step for the almond industry, and processors turned names over with the proviso that the information would be kept confidential by the ABC, and away from CAFF.[9]

The ABC has captured several unanticipated network-strengthening benefits from undertaking the almond PMA, its chief being enhanced credibility in the eyes of the growers who fund it through processing fees. By leading a concrete project that has practical value for growers, the ABC has enhanced its standing among its constituents. The ABC also uses the progress the industry has made in organophosphate reduction in its negotiations with the DPR. In contrast to BIOS, the almond PMA has not raised questions about the character and purpose of UC science, nor about the power dynamics in relationships between growers and extensionists. It has remained focused on helping Farm Advisors develop and extend environmentally responsible practices, and helping growers to deploy these with appeals to economic rationality.

Networks can fall apart. Tensions between CAFF and the ABC have always been present, but during the late 1990s they found that it was to their mutual advantage to collaborate. They cooperated in activities in the space where their objectives overlapped. CAFF was able to achieve its organizational goals by engaging and influencing the ABC, and the Almond Board found it was advantageous in negotiating with regulatory agencies to be identified with an organization widely perceived to be innovative and environmentally responsible. The ABC was always more comfortable with the pro-grower dimension of CAFF than with its environmental advocacy activities. Four years into the five-year partnership, tensions erupted and CAFF was dismissed from the partnership. The proximate cause of this divorce was CAFF's participation in the *Fatal Harvest* project and the "Beyond Organic" campaign that accompanied it. This book, financed by the Foundation for Deep Ecology, attacked industrial agriculture with a broad brush and savage language. Many of its essays are by leading voices for reform in agriculture and are quite thoughtful. These are jumbled with vitriolic denunciations of industrial agriculture and the environmental impacts of agrochemicals. The book included numerous distortions and mis-statements of fact, and Heintz and Farm Advisors described the book as "a hit piece."[10]

CAFF had readily agreed to participate in the "Beyond Organic" marketing campaign launched with the release of the book, and to receive grant monies for this. CAFF did so before the factual inaccuracies and controversial nature of the book came to light, and it became apparent that the book would disrupt their network. Once the book was published, the book exacerbated CAFF's internal tensions between advocacy for change and serving growers. Vigorous debates broke out between staff members, among the directors of CAFF's board, and between staff and directors. CAFF's executive director rebuffed Heintz's request for CAFF to renounce *Fatal Harvest*, and the relationship between their organizations became acrimonious. In September of 2002, the management team of the almond PMA tried again to persuade CAFF to disavow the book, without success. Most of the PMA management team felt that CAFF was attacking them, thus disqualifying their further participation. The ABC terminated CAFF's participation in December of 2002. Knowledge can serve as the foundation for creativity and innovation in partnerships, but arguments about knowledge can fracture them just as easily.

Grower Knowledge versus Expert Knowledge

A comparison of the Randall Island Project and the Prune partnership reveals divergent attitudes about the role of Cooperative Extension in these networks. Grower Doug Hemly and PCA Pat Weddle had prior

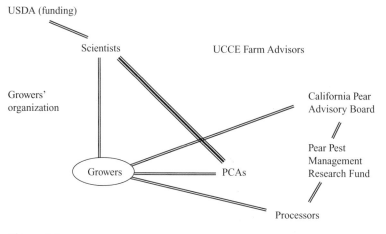

USDA (funding)

Scientists   UCCE Farm Advisors

Growers'   California Pear
organization   Advisory Board

                                                Pear Pest
                                                Management
Growers   PCAs   Research Fund

                          Processors

Figure 6.6
The Randall Island Project network.

relationships with research scientist Welter. Weddle and Hemly persuaded contiguous pear growers to cooperate, but Weddle and Welter led
up small groups of PCAs and scientists to design and conduct the field
research necessary to implement pheromone mating disruption. Thus,
Hemly and his neighbors played the most critical leadership role in
agreeing to try a new technologies and undertake commensurate risks,
and this made possible the essential multi-season relationship between
scientists and PCAs. This project depended on being able to persuade the
codling moth populations to behave as the scientists thought they could.
Support from the California Pear Advisory Board and from the Pear Pest
Management Research Fund made pheromone mating disruption economically feasible. No Farm Advisors participated in this partnership; in
fact, Sacramento County was without a pear Farm Advisor during this
period. Farm Advisors have played roles in other pheromone-based partnerships, but generally much less than in other California partnerships.
Weddle notes that UC Cooperative Extension's professional incentives
now emphasize research for publication at the expense of commercial
implementation. As use of pheromones to disrupt mating has become
routinized, Farm Advisors have a progressively diminishing role in this
type of partnership.

The prune partnership took the opposite tack. Its strategy is to re-
orient the entire prune industry and its knowledge system through

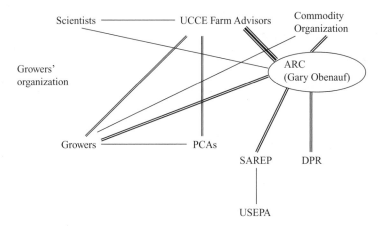

Figure 6.7
The Integrated Prune Farming Practices network.

constructively engaging UCCE. It has been led by Gary Obenauf, a former Farm Advisor who for decades has worked with the California
Prune Board (now legally renamed the California Dried Plum Board). In
1994, he formed his own company, Agricultural Research Consulting
(ARC), and the Dried Plum Board contracts his services as research
director. In part because of this commodity's historical reliance on diazinon, the prune partnership has received funding from SAREP and the
DPR. This partnership illustrates the strengths and shortcomings of making Farm Advisors the center of partnership activities. Prune growers are
the second ranked users of organophosphate dormant sprays, and industry leaders recognized that growers had to develop a different approach
to pest management. Obenauf describes his partnership as an attempt to
blend the best from BIFS with the best of traditional extension. He persuaded participating Farm Advisors to recognize that the scale of change
needed in prune farming could not be done effectively through traditional, large meetings of growers with slide shows, but they resisted the
idea of legitimating grower-generated knowledge in the BIOS model.
Obenauf has invested most of this partnership's resources in developing
and validating a coherent set of practices. He has tried coaxing Farm
Advisors into allowing growers to give testimonials at field meetings,
with uneven success. Management and field meetings are run by Farm
Advisors for Farm Advisors, and there are few genuine opportunities for

growers to contribute their knowledge or leadership. Growers are ostensibly on the management team, but meetings are frequently scheduled at times they cannot attend, and their contributions appear slim. Obenauf believes that the release of the *Integrated Prune Farming Practices Decision Guide* will accelerate the adoption of practices by PCAs and growers.

Local Winegrape Partnerships in Northern California

"Winegrapes are a product of a place," in the words of John Clendenan, a Sonoma County vineyard manager. The geography of winegrapes facilitates and amplifies the successes of agroecological partnerships in this commodity. Winegrape growers are acutely aware that they are paid on the quality of their grapes, and that their location strongly influences the price they receive. More than any other agricultural product, wine is branded by the place the grapes are grown and vinted. Geographic branding creates a structure for coordinated actions among producers, even greater than marketing orders, to add value to their commodity.[11]

Regional cooperation among winegrape growers laid the foundation for their partnerships. Wineries have profited from the geographic branding of grapes and wine, and they too have supported winegrape partnerships more actively than has any other food-processing industry. Not all winegrape regions are created equal, however. Demand for wine has spurred the expansion of winegrape vineyards throughout the state, but most of the profits have been realized in coastal counties.

Unfortunately for the winegrape industry, the regions where they can grow the best grapes also host highly desirable residential real estate, exacerbating urban-agricultural interface tensions. Criticism of the industry on environmental grounds threatens the viability of winegrape production. Winegrape partnerships in coastal counties are the only to explicitly name the public as a target audience for their message, and half of the six partnerships can be found in those counties.

Winegrape partnerships have more fully taken over extension functions than those in other commodities, in part because they are motivated to enhance the environmental credentials of their industry to local critics, and in part because he winegrape industry underwent dramatic expansion at the same time UCCE suffered an extended period of

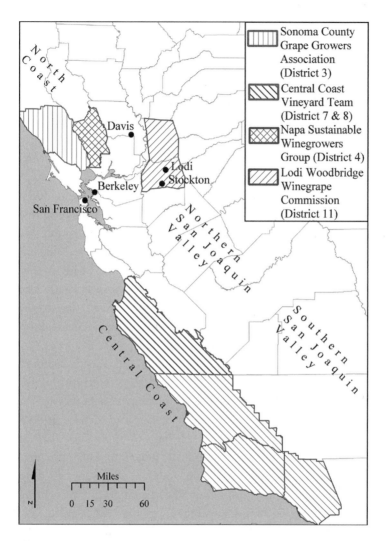

Figure 6.8
Map of California winegrape partnerships.

flat or declining budgets. Together all these factors explain why this commodity hosts six partnerships, more than any other, and four local partnerships, also more than any other commodity.

The Napa Sustainable Winegrowing Group (NSWG) began in 1995 when half a dozen managers for progressive vineyards plus a local resource conservation officer and the county agricultural commissioner recognized that they had developed a significant amount of experience with and knowledge of sustainable practices, but that no forums existed for sharing it. The existing networks of knowledge had been geared toward production or quality, but not sustainability issues. These initial leaders realized that many of them individually had made progress toward more ecologically rational practices, but they could be even more effective if they worked together. One of these leaders described this initial group as "needing a peer group to achieve goals of being able to identify, verify, understand, and promote sustainable practices." The Robert Mondavi Winery hosted initial and many subsequent meetings of NSWG, and provided it early seed money.[12] Other Napa wineries also have supported this partnership.

NSWG appears to be the most independent and least structured partnership in the state, which is likely a result of the personality and tradition of innovation of this region. It does not have "a program," or demonstration vineyards, nor does it prescribe certain practices. It does not formally enroll growers or vineyards. It has not applied for BIFS or PMA or USEPA funding. It promotes a holistic understanding of sustainability as a primary criterion for farming and conducts outreach to support this. Leaders in this group describe the importance of being guided by a sustainable farming philosophy more than other groups.

Partnership activities take two forms: regular monthly meetings and quarterly outreach events. About a dozen growers and vineyard managers participate regularly, discussing specific practices and farming philosophy in a dialectical spirit. The group invests the majority of its outreach effort into several large workshops each year. The titles of their workshops reflect their ambitious, self-reflective vision: "Sustainable Winegrowing in Napa County," "Ecology for a New Millennium," "Water Use and Its Place in Sustainable Farming." This partnership and the Central Coast Vineyard Team are the only that use the term "agro-ecology" in their outreach materials. Beyond these events, NSWG leaders

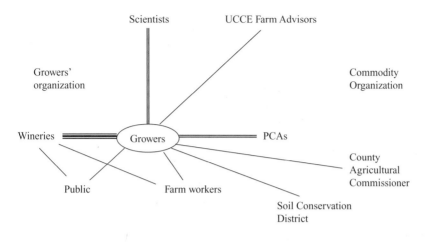

Figure 6.9
The Napa Sustainable Winegrowing Group network.

describe conducting personal conversations outside these workshops with other growers, and inviting them to visit their vineyards to see what practices they have found to be successful. This group is not well known outside Napa County, but it appears to have considerable credibility and some influence there. Its members are acutely attuned to the need to reach out to the public, in part because of the many controversies surrounding Napa winegrape production.[13] This is the only partnership in which a county agricultural commissioner plays an active leadership role. He has a very visible, public office, and it is to his advantage to participate in a group that is working to reduce tensions with the public. This group understands its audience in the broadest terms of any partnership. They explicitly try to reach absentee owners, owner-operators, vineyard managers, pest-control advisors, farmworkers, vintners, regulators, and consumers with their message, connecting sustainability with wine quality.

The social equity dimension of sustainability is addressed more consistently and completely by the Napa partnership than by other groups. The Napa partnership conducts more outreach about and to farm workers than any other partnership.[14] Its leaders have the broadest operational definition of sustainability, yet they have not institutionalized this in a formal structure. They have chosen for their organizational strategy a loose-knit network and may be programmatically weak compared to other partnerships, but they have the broadest and most sophisticated understanding of sustainability.

In 1984, seeking to conduct community outreach and education for their industry, local winegrape growers founded the Sonoma County Grape Growers Association. In the 1990s, land-use battles spread from Napa County to Sonoma County, and in the later part of the decade the Sonoma County Grape Growers Association began to recognize the need for a partnership approach to educating growers.[15] In 1999, the organization hired Nick Frey as its executive director. The association's board hired Frey to help market winegrapes, but they knew they also needed to address their neighbors' concerns about environmental impacts. Frey had "a hunch that people realized that there were some issues out there. . . . We could probably do some things better, and too, we were going to be under closer scrutiny, but I would guess that they underestimated . . . the scrutiny and the timing of it, by like an order of magnitude, because in the fall of '99, it just exploded in the papers, this 'terrible problem' with the vineyards, and we were problems for every kind of issue you could imagine, pesticides just being one of them."

The crisis of public perception provided a powerful motivation for these growers to collectively demonstrate their environmental values to neighbors and consumers. Many have felt wronged by complaints about their vineyard practices, and are motivated to disprove critics. The grower John Clendenan said: "Traditionally the farmer had the full say over what happened on his land, and that picture's changing really fast. And so it was very important to us to present a positive light on what we did. We were perceived as the 'green desert,' and then there were particular hot points that we've started to be attacked, viciously attacked on, usually sprays, methyl bromide use, certain pesticide use."

Frey launched an IPM committee in 2000; later it was renamed the Sustainable Practices Committee. The association has been unsuccessful in obtaining a BIFS grant but has received FQPA monies from the USEPA. It has compensated for its limited funding by facilitating active participation by many growers and PCAs. Because the Sonoma County Grape Growers Association has had to rely on social resources, this partnership has strengthened the capacity of its parent organization, and added value to the organization in the eyes of its member growers.

Leading growers in the Sonoma winegrape partnership have been particularly active in outreach to other, less environmentally oriented

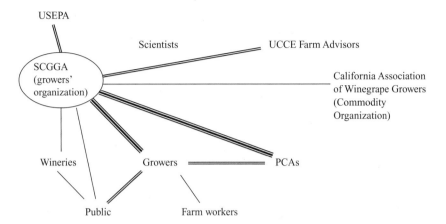

Figure 6.10
The Sonoma County Grape Growers Association Sustainable Practices Committee network.

growers, including those who are not members. The profile of grape growers in Sonoma has changed dramatically since the 1970s. Growers here seem to fall into one of three groups: Long-time growers, who may have grown grapes conventionally for a long time, or who might have expanded their operations into winegrapes; urban refugees who are farming for the first time, and chose winegrapes because of their cachet; and third, a group of progressive growers who have been actively leading the local winegrape community to improve their practices and reach out to the non-agricultural public. Leading growers actively recruit the conventional and the new growers to attend the meetings and purposively engage them. One described the partnership as trying to "sweep in more of the people back in the corner" because it is "nice to raise their consciousness a little bit." Leading members of the Sonoma County Grape Growers Association are motivated to engage conventional growers because they perceive the importance of enhancing the public image of winegrape growing in their county.

The partnership established four demonstration vineyards in 2000, one in each of the major sub-regions in the county. The owners of these vineyards were association board members, and some of the most progressive winegrape growers in the county. Monthly grower meetings are held in these four vineyards throughout the growing season, led by Nicki Frey and Laura Breyer. As a well-respected local independent PCA,

Breyer plays a central role in the outreach efforts of this partnership. She explains information to growers in a non-threatening, non-technical way, and comes across as something of a low-key biologist/mentor. She convenes PCA breakfasts monthly during the growing season, and helps identify pest trends, approaches to monitoring, and new products. This is the only partnership to use an independent PCA to conduct outreach to other PCAs, and this has been a very effective strategy for using the USEPA funds it has received. Members of the management team describe their local field days with the following terms: learning by interaction; providing positive peer pressure; motivating growers to look good in front of the group; bringing new eyes on old problems; and raising the bar for other growers. These events provide a space for growers to engage each other.

In both Napa and Sonoma, collective action was necessary to address crises in public perception. Leading growers and vineyard managers recognized that the winegrape industry is vulnerable to public perception, and that they could only make progress in addressing this problem by changing the image collectively. Winegrape growers think of themselves as producing a crop in a particular place, and to address the crisis in public perception, the collective producers had to take action to address the concerns of local residents. Both Northern California winegrape partnerships provide platforms for leading growers to enroll others in their vision of agriculture, but they have chosen divergent network designs: the Napa partnership is an informal association, and the Sonoma County Winegrape Growers Association has expanded its service to its members, and leveraged increased growers' participation for its mission.

Local versus Statewide Winegrape Networks

In the early 1990s, the Robert Mondavi Winery launched an initiative to enhance the quality of wines made in the Central Coast region. This evolved into the Central Coast Vineyard Team. Even though this growing region is geographically large and spatially dispersed, most vineyard management decisions are concentrated in a relatively small number of growers, vineyard managers, and industry leaders who have worked together over several years (these are all represented in the sociogram by "growers"). Wineries exert more control over this region than other pre-

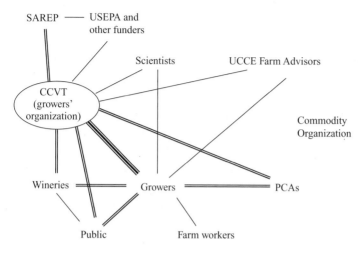

Figure 6.11
The Central Coast Vineyard Team network.

mium production districts, and they have supported the Central Coast Vineyard Team in part to protect their public reputation.[16] The Positive Points System appealed to participants because it provided a structure for learning from each other, but it also gave them a tool for communicating to the public.

The same environmental controversies that dogged the North Coast began to find expression in the Central Coast region in the mid 1990s and similarly stimulated vineyard owners, managers, and employees to debate the operational meaning of sustainability. The winegrape industry in this region is sufficiently concentrated that its leaders knew each other, and the process of creating the Positive Points System intensified their social network.

Members of the group have addressed tensions between creating exclusively scientifically objective criteria and the need to represent themselves to the public. The team uses the PPS to represent themselves to the public. It has received funding directly from the USEPA and indirectly from that agency through SAREP. Team leaders feel confident that the PPS measures sustainability at least as well as it measures any other criteria, and they welcome scrutiny by their critics. The PPS has given these growers additional confidence in being able to face vineyard critics in an open way. The objectivity of the criteria is useful for growers scrutinizing each other, but also for presenting themselves to the public. One Central Coast Vineyard Team leader said: "In the very beginning, we started inviting in opposition groups, opposition point of views, and there was never any hesitation on our part. I think that the group felt pretty righteous about our, where we were, and we didn't have anything to hide, that given an opportunity to tell our story, it would be a compelling story and that it was our responsibility to do that." The team regularly holds forums and vineyard tours in the various Central Coastal counties to respond to community concerns, including an annual field day for elected officials.

The California Association of Winegrape Growers (CAWG), a statewide winegrape commodity organization, followed the success of the Lodi partnership and the three local partnerships with interest. All winegrape growers are required to support CAWG, but it also worked closely with the Sonoma partnership and with the Central Coast Vineyard Team. Winegrape growers have to actively manage public opinion about the environmental impacts of their vineyards because premium winegrape vineyards are more consistently inter-digitated with suburbs in coastal counties than any other crop. Karen Ross, CAWG's director, recognized the multiple benefits that partnerships provided, and she had participated in Wine Vision, a national wine industry initiative to enhance their economic vitality into the next century. CAWG leaders recognized the opportunity to serve its members by applying for a PMA grant, so they consulted with local growers groups and determined that sulfur drift and persistent, pre-emergent herbicide use were the thorniest issues facing them.[17]

State legislators had expressed concerns about sulfur drift to the DPR, and this favored the funding of the winegrape industry to conduct outreach to its growers. CAWG received a PMA grant to fund a statewide partnership to address these issues, and Ross designed it to be a partnership of partnerships. The Winegrape PMA provided educational, outreach, and media resources for local groups, facilitated the sharing of educational resources between regions, and worked to strengthen the capacity of regional groups that did not have their own partnership. This partnership did not aspire to the integrated farming systems ideal, and its leaders made no pretense of this. It addressed two top regulatory problems for the winegrape industry. This partnership did not include any

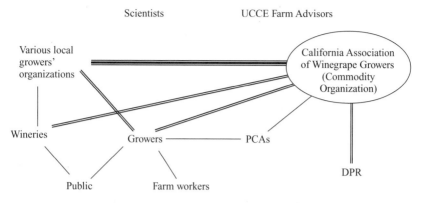

Figure 6.12
The Winegrape Pest Management Alliance network.

research component, and very little participation on the part of UC scientists or extensionists.

Some agricultural leaders outside the winegrape industry have criticized this partnership for being only focused on the reduction of two agrochemicals, but leaders of other, local winegrape partnerships have been quick to defend it. They point out that regions that wanted to take an integrated farming system approach have done so, and this did not need to be duplicated, even if it were possible at a statewide level. The winegrape PMA has served to cross-pollinate some of the outreach ideas from the more active regionally based groups to areas without formal partnerships. It operated more as a network of (local) partnerships, or an institutional partnership. Individual growers have participated, but it has not been led by them, as have local partnerships. It laid the groundwork for the Code of Sustainable Winegrowing (described in the previous chapter).

Winegrapes are California's most valuable plant crop, worth $1.8 billion in 2000. This commodity's value and acreage grew dramatically during the 1990s, so developing in partnership activities is unsurprising. These four cases show, however, that winegrape growers used the agroecological partnership model to achieve multiple goals, and adapted it creatively to take advantage of their existing social networks. More than any other group of growers, they used the partnership model to persuade the public of their environmental initiatives. They fused the notion of wine and environmental quality, and made a virtue out of necessity.[18]

Conclusion

Pat Weddle observes that new technologies such as synthetic pheromones are impacting farming faster than its traditional agriculture science support institutions can respond. Agricultural technologies—many of them designed to replace organophosphates and other problematic agrochemicals—are being commercialized before they can be researched. Agriculture may or may not be saved by these technologies, but it certainly is being assaulted by them. Some of them are legitimate, but without field trials and knowledge networks, how can growers know whether or not they offer a real economic benefit? Agroecological partnerships are a chief strategy for connecting the extra level of expert knowledge with the practical knowledge of growers.

Previous chapters have described partnership participants and the kinds of practices deployed. This chapter has illustrated the dynamics of knowledge exchange, and allowed us to compare different strategies for configuring these networks. All draw from a mix of basic and applied knowledge, and all address environmental impacts using agroecological strategies and practices. Each commodity has its distinct history of social relations, but sustained efforts require the participation of all categories of participants. All have critical roles in circulating knowledge through Latour's model of science. As this chapter has demonstrated, there is no one right way to organize these networks. They emerge to fill gaps in agroecological knowledge systems. Networks evolve through stages, just as partnerships shape commodity knowledge systems in stages. (See chapter 3.)

Agroecological partnerships draw from and apply knowledge generated by California's pre-existing science networks. Partnership networks articulate basic, laboratory-derived knowledge with field application. Pheromone-based partnerships in particular require sophisticated technical skills, and the economic advantage of these technologies must be clear enough to justify the additional expense of this expertise. Partnerships build on existing networks and extend them by generating and sharing knowledge. Their configurations manifest theories of change held by partnership leaders by making one or more sets of actors the focus of additional knowledge support. Network analysis explains how partnerships engage actors in the multiplicity of their relationships and how they provide incentives for their participation, or lack thereof. Actors in California agriculture operate in many different networks simultaneously, and successful partnerships have been able to enhance multiple relationships among these actors. Some growers, extensionists, and institutions within existing networks found agroecological partnerships to be a powerful vehicle for them to be able to express their vision for a different kind of relationship between nature and society, guided by an alternative agriculture.

Organizations hosting partnerships substantively shape them through the primary dyadic relationship. These types of partnerships are complex undertakings with multiple goals, components, and participants, all of whom necessarily bring some of their own interests to the shared project. Thus, partnerships are networks that demand political skills to be able to manage their stated and explicit goals, as well as their assumed and competing goals. Networks can become stronger by taking on multi-faceted projects, but they can also become more difficult to manage than centrally organized institutions. The progress that these some partnerships have been able to make in helping growers re-orient their production practices is a tribute to their leading growers, but also the creativity and organization-building skills demonstrated by partnership leaders.

Most agroecological partnerships struggle to re-shape the larger market and regulatory institutions to favor partnership practices. Processing and marketing companies express verbal, but generally limited financial support for partnership activities, with winery funding of winegrape partnerships the important and notable exception. Commodities with successful partnerships owe their accomplishments in part to their ability to circulate knowledge broadly among all the principals, which the next chapter will discuss.

7

Circulating Agroecology

Growing Grapes

Franciscan Friars brought European grapevines to California. Junipero Serra apparently planted them at the San Diego mission in 1769, and the Friars carried cuttings up along El Camino Real as far north as Sonoma to plant vineyards for fresh grapes and winemaking. The Mission varietal is a hardy vine: its fruit has high sugar content but lacks flavors and color. The Franciscans were either lucky or clever because the Mission yields a good eating grape, a passable raisin, and a juice that can be made into (sweet) wine.

Grapevines are malleable plants. During California's Mexican period, Los Angeles grape growers began to ship fresh table grapes to Northern California, and this increased dramatically during the Gold Rush. The Davis-area growers R. B. Blowers and G. G. Briggs were the first to plant the Emperor varietal for table grapes. They were the first to pack grapes on trains to ship them to the eastern states in 1869, creating demand for transcontinental ice-cooled refrigerator railcars. Growers around the state began planting dedicated varietals for raisin production in the 1860s and the 1870s. Raisin production eventually settled in to the Fresno area because it offers the ideal climatic conditions: grapes grow well there, and its dependably rain-free Septembers dry them in the field. This region now hosts the greatest concentration of raisin production in the world.

One Gold Rush-era entrepreneur in particular recognized that California's varied landscapes could support a diversity of grape varietals, and resulting fine wines. Colonel Agoston Haraszthy emigrated from Hungary and was the leading promoter of viticulture in this state during the first two decades of statehood. He toured Europe to survey the state of winegrape growing and winemaking, and brought back some 200,000 cuttings to plant in his Sonoma County vineyard. He relentlessly promoted scientific viticulture and experimenting with varietals to determine which did best in particular production regions, demonstrating the diversity of farming conditions under which winegrapes could be cultivated.

Prohibition dealt a harsh blow to winemaking, setting back wine quality for decades. The year 1976 marked California wine quality coming of age when blind taste tests revealed Napa wines to be superior to French in Paris by French experts. The "Judgment at Paris" transformed the California wine industry because winegrape growers and wineries realized they could make far more money by pursuing the quality market.

In *Bottled Poetry*, Jim Lapsley explained how Napa Valley wineries and grape growers undertook quality improvements with a near-religious zealotry starting in the 1950s.[1] California winegrape growers and wineries have worked together for decades to improve wine quality and differentiate their product in a highly segmented market. They have profited from improving the quality of their product, branding their place of production, and conveying knowledge about production conditions through labels to consumers. The California wine industry's ability to capture economic benefits from the premium market was made possible by the circulation of better knowledge between scientists, growers and wineries, and growers' ability to feed that information back into their vineyard practices.[2]

Four main factors determine superior wine: environmental conditions of production, varietal selection, vine management, and winemaking skills. The first has contributed to the importance of winegrape growing districts, described in the previous chapter. Once a vineyard is planted, there is precious little one can do to influence soils or climate. The French word "terroir" communicates their importance to quality, but it does not translate easily to American viticulture.[3] The acid and flavor content of grapes in particular depends on temperature fluctuations caused by warm days and cool nights just prior to harvest.

The winegrape grower's chief responsibility in this system of production is to manage the grapes so they can be harvested at their peak. Premium wineries measure quality by ripeness, flavors (acids, etc.), Ph, and degrees Brix (sugar content), although superior quality requires the grower to balance these consistently. Premium wineries and winegrape growers work together throughout the season to ensure a vineyard ripens simultaneously. Nothing compromises wine quality more than under- or over-ripe grapes.

Table and raisin grape growers in the central and southern San Joaquin Valley have oriented their farm management much like most of US agriculture: measuring success in tons/acre, boosted by high levels of chemical fertilizers and irrigation. Kern, Tulare, and Fresno Counties have plenty of heat units, but autumn nights do not cool sufficiently to boost acids and flavors. As a result, vines here grow vigorously and have a heavy grape set, but they produce inferior wines. Wine made from these grapes is an undifferentiated commodity, and it is processed and marketed much the same as (unfermented) grape juice.

Researchers thoroughly documented the production and environmental problems associated with excess nitrogen in vineyards: over-application of nitrogen to grape vines can result in excessive vigor, which stimulates canopy growth shading the fruit, creates a favorable microclimate for pests and disease, and decreases fruit quality. High levels of nitrogen in plants attract leaf-eating pests regardless of the crop, requiring pesticide treatments. The southern San Joaquin Valley averages of 10 tons of grapes per acre, with some vineyards producing as high as 15 tons per acre.

In contrast, Napa growers learned that moderate vine growth produces better quality winegrapes. The two primary strategies they developed with the UC Davis Department of Viticulture and Enology were canopy and input management. Growers and farmworkers manage the canopy by pruning, chiefly during the winter, but in some cases remove excessive leaves during the growing season. This increases the sunlight on the grape clusters, which in turn increases grape quality and reduces the risk of powdery mildew disease. Winegrape growers learned that using less fertilizer and water would increase the intensity of flavors and value of the grapes, but at the cost of reduced crop tonnage. In some cases, winery field staff may ask growers to "drop" or thin grape clusters, and growers agree to this because they anticipate a higher per ton payment or maintaining a profitable relationship with the winery. Yields under three tons per acre are common in the Napa region, but the price per ton here is ten to fifteen times the average for winegrapes in the southern San Joaquin Valley.[4]

California winegrape growers in general knew that they could make more money by producing fewer grapes if they could sell them at a premium, but as consumer demand drove wine prices ever higher, a few regions organized themselves to capture economic opportunities in the intermediate segments of the wine market. The efforts by the Lodi region grape growers make even more sense in this light. Throughout its first hundred years of winegrape production (1880s–1980s), Lodi winegrape growers grew the same grape varietals as their highly productive neighbors to the south, and judged their crop by the same criteria: on a tons/acre basis. Lodi wines were unremarkable and undifferentiated. They had no reputation, so wineries did not place the Lodi name on their bottles.

As the 1990s began, growers in the Lodi region faced a decision. They could try to follow behind Napa as it climbed the quality ladder. This would entail risk as they grafted over their vines to better wine varietals and tried to improve the quality of their grapes. But continuing down the path of growing undifferentiated, unremarkable winegrapes entailed risks as well, since their production costs were high and getting higher, and winegrape growers from other countries with lower production costs were aiming at mid-level price points. Other California winegrape growing regions also had an eye to follow Napa. As the Lodi grower Randy Lange put it, "it was eat or get eaten."

The Lodi Woodbridge Winegrape Commission and its member growers have done a remarkable job of adding value to its winegrape crop. The annual average price per ton in the district, roughly twice that of its down-valley neighbors, only tells a part of the story. Since the launching of the commission, the district's winegrape acreage has doubled to 80,000 acres while the number of growers has only grown by 18 percent. The number of wineries in the district has grown from eight to fifty. The number of wines with Lodi on the label has jumped from four to 150. Large wine corporations, including Mondavi, Gallo, and Canandaigua, buy about 70 percent of Lodi's grapes, and they are putting the Lodi name on their labels and charging more money for these wines. The other 30 percent of the grapes are going to smaller wineries, many of which now pay more for Lodi grapes. Lodi growers are finding it easier to get contracts with wineries, easier to stabilize their relationships with wineries, and the long-term trend in their grape prices is stable or rising. Most American agriculture operates under the assumption that increasing yields is the only way to maintain profitability. California winegrape growers in general and Lodi growers in particular have developed a contrary strategy that has greater promise for the future of American farming.

Lodi growers and their commission undertook quality enhancement initiatives simultaneously in product and production. They were fortunate to grow a crop that lends itself more easily to adding value through quality improvement, and they invested considerably in re-orienting their production toward quality. Their location just east of the Sacramento Delta means they face continuous, close scrutiny of their practices and impacts on water quality, but they were able to take advantage of the cool breezes to grow improved quality winegrapes. The growers, Cliff Ohmart, and the commission staff have now addressed sustainability in both environmental and economic terms for more than 12 years. They developed the longest running, most comprehensive social learning effort in recent California agriculture, circulating agroecological knowledge through farm to growers to consultants to their organization, and ultimately to the public. Linking environmental and product quality has been essential to the Lodi partnership's success, but their most important strategy has been to facilitate active exchange of knowledge among all participants, including regulatory agencies, wineries, consumers, and the public at large. They have done this by building a highly sophisticated network that fosters social learning.

This chapter describes how partnerships have coupled knowledge about agroecological strategies and practices with economic incentives. This book has documented how alternative, agroecological initiatives can make genuine progress toward environmental resource conservation goals, more than any other approach. But as Latour reminds us, this kind of alternative science must be well represented to the public, and the public's support must ultimately be enrolled to support this kind of hybrid economic/ecologic project. The Lodi winegrape has developed the agroecological partnership model more fully than any other effort. Winegrape growers have several distinct advantages that growers of other crops lack, but their model of collaboration and practical, applied science in service may help other industries develop strategies. Indeed, as commodities that have benefited from agroecological partnerships, the almond and pear industries have developed some of the same strategies of cooperative knowledge exchange and adding value, and representing these to the public, not merely public agencies. These initiatives may be able to contribute to the broader economic crises in agriculture.

Circulating Knowledge in the Lodi Winegrape Partnership

More than any other, the Lodi winegrape partnership has actively circulated knowledge through all five of Latour's loops. As a commodity, winegrapes have several traits that make partnership initiatives easier. Growers participating in other partnerships complain that their crops do not have the same advantages as do winegrapes, with remarks like "winegrapes are different." The winegrape industry has developed its partnerships more than other commodities, but the social and economic forces pressing against California agriculture do so regardless of crop. Every other commodity will have to develop strategies to add value to their crops if they are going to survive in the US economy. The investment of effort and resources like that of the Lodi partnership will be necessary to ensure agriculture's future in rapidly urbanizing regions, such as in most coastal states. The Lodi partnership has developed a full toolbox which can be adapted by other agroecological initiatives.

Growers created the Lodi Woodbridge Winegrape Commission to improve practices, enhance the collective quality of their grapes, and improve their protection of environmental resources. Leading growers decided not to do this apart from their neighbors, but rather enroll all growers in a cooperative project. As soon as the region's growers voted the commission into existence in 1992, it established three main goals:

differentiate Lodi's winegrapes and wine in the market
fund a regionally specific research program to benefit local growers
encourage the use of environmentally friendly farming practices.

Note that all three goals are dependent on the sharing of knowledge. The third goal came about when the grower John Kautz persuaded the new executive director of the commission, Mark Chandler, to integrate an IPM component into the new organization's efforts. Leading growers readily agreed to it because they had invested their time and treasure in the creation of the commission, and saw IPM as a way to better manage their inputs and add value to their crop. Since its inception, the commission itself (with monies from growers and agencies) has funded nineteen region-specific research projects with more than $600,000. These include: biological control of insect pests; the relationship between cover crops, nutrient management and IPM; varietal and trellis research; and irrigation management.[5]

Cliff Ohmart describes the commission's efforts in four stages: grower IPM outreach (1992–1995), the BIFS grant (1996–1998), area-wide implementation of the *Lodi-Woodbridge Winegrowers Workbook* (1998–2003), and the Lodi Rules for Winegrowing (2003+).[6] Each stage added another degree of sophistication to their network of relationships, another layer of knowledge shared about how to integrate "sustainability," agroecological practices, and improved product value. Initially the commission and Ohmart circulated knowledge between nature (the vineyard), scientific colleagues, clients (growers), and scientific content, but once they began to use the workbook and think more seriously about using their practices to help market their products, they expanded into the "public representation" loop, more so than other partnerships.

The first stage was conceived as an effort to familiarize growers and their PCAs with IPM techniques, using a variety of field days and printed educational materials. UC IPM Director Frank Zalom successfully obtained a $150,000 Kellogg Foundation grant to support this stage. It sought to fulfill the original vision of IPM developed by UC Berkeley scientists in the 1950s, with specific goals to reduce reliance on synthetic chemicals, develop a district-wide IPM program, and encourage IPM techniques. They achieved these goals by publicizing local agroecological knowledge at breakfast meetings and field days.

The second stage began when the commission secured one of the first two BIFS grants from SAREP in 1995, and hired Ohmart to run it. Ohmart implemented the BIFS program guided by his experience educating growers as an independent PCA and his observation of BIOS's grower-to-grower outreach model. He worked with 43 growers to designate 60 vineyard blocks covering 2370 acres for intensive monitoring and evaluating alternative management strategies, and to facilitate them sharing what they learned with all 700 growers in the region. Ohmart's overall goal was to help growers recognize the importance of enhancing biological processes in their farming operation. The BIOS vision of a grower-to-grower outreach model was fully institutionalized in the Lodi Woodbridge Winegrape Commission.[7]

The three basic components to the BIFS program were grower outreach, field implementation, and evaluation. Outreach included neighborhood grower meetings, field days, breakfast meetings, half-day research seminars, and newsletters. Ohmart worked with leading growers to demonstrate field implementation, to operate BIFS demonstration vineyards, and to host field days and other events. These growers managed 40 percent of the district's winegrape acres, and have very high credibility in the region. Many of them had campaigned to create the commission, wanted to learn more though this partnership and encouraged other growers to learn about these practices. During the late 1990s, growers already familiar with IPM approaches applied those techniques more consistently throughout their vineyards, and those new to IPM were repeatedly encouraged to monitor their vineyards and make pesticide decisions based on data about economic thresholds.

The 14 PCAs enrolled in BIFS monitored half the acreage in the district, and even affiliated PCAs came to recognize that they were helping add value to the growers' crops. Several of the affiliated PCAs in the district realized this was the direction the PCA industry was heading. The affiliated PCA Steve Quashnick said: "That train [of change in agriculture] is leaving the station, and I can either get on it or watch it leave."[8] He says he now questions every pesticide application. Even though the commission's efforts challenged his established business practices, he came to recognize and value their innovative character, and he realized that if he did a good job serving the needs of growers, he would have an economic future.

Buoyed by the success of the BIFS program, in 1999 Ohmart began a third stage, based on the development and implementation of their workbook, described in chapter five. Ohmart does not simply pass out the workbook, but rather presents it in half day workshops to help growers understand how to put its recommendations into action. This self-assessment describes and affirms good farming practices being done, identifies farming practices that are of concern from an environmental or wine quality perspective, and develops an action plan, and a time table for making measurable improvement. Between 2000 and 2003, Ohmart presented 48 workshops in the district. Nearly 300 Lodi growers have taken the workshop once, and one-third of them have taken it twice. This represents almost all of the active full-time growers in the district. These growers manage 68,000 of the 80,000 acres in the district.

The commission recently began the fourth stage of its program, the Lodi Rules for Winegrowing. Based on its prior work developing a farming system approach and the Lodi workbook, the commission's goal is to define, measure, and implement "sustainability," with the intent of being able to demonstrate this to wineries and the public. The Lodi Rules are the first peer-reviewed standards for winegrape growing, and they are certified by an independent third party, Protected Harvest. The standards were developed jointly by growers, scientists, academics, and environmentalists, and are customized to the specific ecological conditions of Lodi. The rules have two components: sustainable winegrowing standards, and a Pesticide Environmental Assessment System. The 75 farming standards are organized into six chapters: ecosystem management; education, training and team building; soil management; water management; vineyard establishment; and pest management. The Pesticide Environmental Assessment System measures the collective environmental impact of pesticides used in a vineyard. To qualify for certification a grower must score more than 50 percent in each chapter of farming standards, and not exceed a determined threshold of pesticide environmental impact. These winegrape growers will use the Lodi Rules to represent their environmental stewardship, and have enrolled Protected Harvest to verify this.

Cliff Ohmart has been central to the success of Lodi's partnership activities. He alternates roles as scientist, teacher, dentist, and ambassador, all bound together by his circulation of knowledge among growers, PCAs, scientists, and regulatory agencies. His PhD in entomology from UC Berkeley helped him direct research funded by the commission, but when speaking to growers he invokes his experience as an independent PCA. Before working for the commission, he developed a computer program to help him organize and then analyze pest monitoring data from his clients' orchards. He insists to Lodi growers that once he invested the time and money to set up a data management system his job became easier, and that he made better decisions as a result. Once he began working with the BIFS program, he developed a relational database that generated a complete data sheet for participating growers. During the workbook phase, he improved the database so that it stores and presents multi-season insect monitoring data, and integrates it with irrigation, fertility, and soil information, offering growers a complete set of vineyard indicators. He regularly extols this "data-driven approach."

Ohmart presents painful but necessary information about pesticide use. Because leading growers and the commission have created a social expectation that pesticide reduction is genuine goal, the original BIFS growers in this district continue to annually review as a group their pesticide use on their original BIFS blocks, with the growers' names attached. These data are generally arrayed in something of a bell curve, with some growers using virtually no pesticides and a few growers using much more than the average. This degree of sharing documentation of pesticide use with growers' names attached is highly unusual in agriculture. Just as dentists repeat the instruction to brush and floss, Ohmart repeatedly stresses the importance of conducting monitoring and evaluating economic damage thresholds before turning to pesticides.

At a meeting in March of 2004, Ohmart and his associate Chris Storm shared (anonymous) data comparing pest counts with pesticide use, and this too revealed a few growers had sprayed far below any economic threshold. In one case, an insecticide was applied when just one leaf hopper was found. Storm also discussed a situation in which one grower sprayed for leaf hopper pests when only the invulnerable adults were present, not the vulnerable eggs, and at a time when only beneficial insects were likely killed. The grower had to spray again later in the season. With the active consent of growers, Ohmart and Storm made visible bad farming decisions being made in the district. They do not shame individuals, but establish a tone that presses against irrational pesticide use with statements like: "this grower was throwing his money away."

Ohmart repeatedly offers affirmation to the growers for the progress they have made but he also travels throughout California as an ambassador for sustainability, both within the winegrape industry and to agriculture in general. He gave 26 talks about the commission's approach to sustainable farming 2002–2003, and his efforts to promote this approach are broadly perceived to be integral to the overall commission's success. Ohmart in turn gives credit to the district's growers, and says that they have been able to achieve what Napa and Sonoma have not because his growers have remained unified (Sonoma's efforts to organize a marketing order were not successful, and Napa growers have not brought it to a vote). Rather than resisting scrutiny, they want to look at data. Unlike most growers, they want to circulate knowledge.

Lodi growers have actively engaged partnership activities. Leading growers in the Lodi Woodbridge Winegrape Commission have carried forward the entrepreneurial tradition of California agricultural capitalists, but incorporating improved quality and environmental concerns into their business model. Some of the larger growers have planted new vineyards while simultaneously buying out or assuming management of lands belonging to others. John Ledbetter began with 2,500 acres in the late 1970s. In the early 1990s, his family-owned corporation managed 4,000 acres in Lodi and Sonoma, and it now manages 11,000 acres of vineyards throughout the state. John Kautz started with 12 acres in 1948, but his family farming corporation now manages 5,000, plus a winery in the Sierra foothill town of Murphy. Randy Lange inherited 150 acres in the late 1970s, now owns 1,500, and manages a total of 6,500. Lange sees himself as born into a transitional generation of farming in California, a "back to the future" period. He does not want to be organic, nor does he see himself able to farm without agrochemicals, but he does want to run his operations with the least amount of disturbance in his vineyard and the least impact possible on resources. This approach flows from his expressed belief that this is the right thing to do. These growers gambled that expanding their operation and improving the marketability of their grapes would be successful. They constitute a small number of growers, but when they speak in favor of the commission's programs, they have great credibility among their peers.

Lange believes that Lodi growers have become accustomed to sharing sensitive information, and they are now more interested in improving as growers. The commission's partnership activities have provided critical knowledge but also social support for incorporating "sustainability" into how they evaluate their farming systems. Lange described how "scary" it was at the beginning to evaluate his operation—knowing that he would be sharing some of that information with others—but insists this is the critical beginning for making a transition toward sustainability. He believes that when Lodi growers have shared information they have helped each other avoid mistakes, raised "awareness in the district," and collectively improved the quality of their wines and reputation. But all this is predicated on a shift in the "mindshift of the grower community." He describes this "mindshift" as growers' asking themselves, before applying any agrochemicals, "Why am I doing it? When should I do it? How do I do it?" Lange sees this as critical if Lodi growers are going to claim they are producing a sustainable agricultural product.

Surprisingly, no studies of Lodi's pesticide use have yet been published.[9] The commission surveyed its growers at the end of the BIFS grant in 1998 and in 2003.[10] The survey's results show how growers report the impact of partnership activities: they monitor more frequently, more thoroughly, and spend more time doing so. They also show that in addition to using monitoring, economic thresholds, and reduced rates of pesticides, many growers were using agroecological cultural controls (e.g., irrigation, habitat and dust management, leaf pulling, modified vineyard trellising, beneficial insect releases).

In 1998, the most accessible outreach efforts (newsletters, neighborhood grower meetings) reached the majority of growers in the district. The smaller percentage of growers dedicated enough to attend more time consuming events (research seminars) had a high rate of satisfaction. But the Lodi partnership owes its success chiefly to the social relations it has facilitated through these four stages of partnership activities. In 1998, growers ranked the six of their seven most useful sources of information

Table 7.1
Most important sources of knowledge for Lodi winegrape growers, ranked. Sources: Ohmart 1998, Dlott 2004.

1998	2003
1. PCAs	1. Farm Advisors
2. Other growers	2. PCAs
3. Farm Advisors	3. Commission IPM staff (i.e., Ohmart and Storm)
4. Field crew	4. UC IPM manual
5. Commission and BIFS newsletters	5. Commission field days
6. Winery personnel	6. Commission newsletters
7. Commission BIFS meetings	7. Lodi breakfast meetings

as people, not documents or web pages. The human proportion of knowledge sources slipped some in the 2003 survey, probably because growers had learned about and experimented with most of the new practices. These responses show how important partnership activities have been to the commission's success.

Lodi winegrape growers perceive the fusion of environmental and winegrape quality as their most important economic opportunity. They ranked these as major management practice opportunities:

improving winegrape quality (more than 80 percent)
supporting efforts to increase US wine consumption (more than 75 percent)
further strengthening their regional identity (more than 70 percent)
adding new, available technologies (more than 50 percent).

Increasing yields per acre is the most common opportunity perceived by most US growers. This is seventh on the list for Lodi growers, identified by fewer than 20 percent. Lodi growers have undergone a major shift in mindset. Quality improvement is a long-term driver for sustaining agriculture and improving its environmental performance.

The Lodi Woodbridge Winegrape Commission's efforts have been successful because its members have supported and participated in a social system that facilitates knowledge exchange, and because they have been able to hitch agroecological knowledge to economically useful knowledge about quality. The growers have repeatedly affirmed their willingness to "self-tax" to support the commission's efforts by majorities of more than 90 percent. They continue to invest in this partnership, and they find useful its services, both in research/extension and marketing. This social structure circulates applied ecological knowledge throughout all five loops of Latour's model. The Lodi partnership has continued because the commission provides continuity and cohesion. As environmentally responsible as some of these growers are, they have been able to profit from improving their collective reputation, which is something few individuals can do alone. The BIFS grant was an essential part of progressing toward partnership goals, but the commission was able to parlay this into an on-going process of social learning.

The California winegrape industry in general is recognizing the value of communicating to the public their efforts to protect environmental resources. This commodity faces the greatest danger of negative public perception because the best grapes are grown where public environmental concerns are quite high. The Napa and Central Coast winegrape partnerships in particular have publicized their progress in reducing environmental impacts. Geographic branding of wines positions the industry to market "sustainability" directly to the public, but also increases the risks associated with any potential environmental or food safety controversies. Winegrape industry leaders are working to convert this vulnerability into an economic opportunity. The winegrape industry has a long history of close cooperation with the wine industry, and wine, the product consumers see, has few cosmetic issues. In addition, wineries have convinced consumers that wine should "tell a story," and "sustainability" is a popular story. Wine consumers are well educated, affluent and generally concerned about environmental protection.[11] The winegrape industry has manipulated more of its advantages for marketing "sustainability" than any other commodity, and thus circulated agroecological knowledge to the public.[12] Advocates of sustainable agriculture laud this progress, but also note that wine is a luxury crop, without nutritional benefits.

Winegrape partnership growers understand their need to represent their industry to the public, and the way partnerships can accord them legitimacy. Participants in all four local winegrape partnerships described tensions between those who want to direct resources toward public representation versus those who want to better verify the positive impact partnership practices have. One Napa partnership grower said there has to be "substance behind the perception of sustainability." Several participants affirmed the importance of backing up their claims about selling the "story of sustainability" with objective information, and acknowledged their efforts could backfire if not substantiated.

Leadership in the winegrape industry is diffused among its several regions. The statewide commodity organization, the California Association of Winegrape Growers, has worked with growers and regional organizations to balance the impulses of competition and cooperation. Karen Ross, CAWG's executive director, explains this cooperation with several factors. Many in the industry recognized the value of local, peer-to-peer initiatives like Lodi's, and a collective realization that that the Australian wine industry's stated intent to become the dominant global supplier within 25 years ("Australia 2020"). Ross describes these as "pre-competitive" issues for the California winegrape industry. California winegrape growers do not receive federal crop subsidies, and recognize they have to cooperate to compete on a global stage for consumers.

The Wine Institute, the California winery trade group, was conducting a strategic planning process during the time several winegrape partnerships began to attract attention, and the Institute recognized the marketing value of becoming a leader in "sustainable" practices, so this was included in "Wine Vision," along with encouraging American consumption of wine, and promoting US wines in the global market. From Wine Vision, the institute partnered with CAWG to develop the Code of Sustainable Winegrowing Practices (described in chapter 5). The Code is designed for a statewide audience, but also for adaptation to local regions. Its primary audience consists of vineyard managers and winemakers, but CAWG and the institute are trying to appeal to all the people and institutions (neighbors, regulatory agencies, input suppliers, customers) with whom the industry interacts. Outreach with the Code is through half-day workshops, in which vineyard and winery managers work through the book, learn about the criteria of sustainability promoted by the Code, and assess their own operations in light of this.

The Code builds on the strengths of previous partnerships, but marks a new phase in the agroecological partnership model. It is the most sophisticated analytical tool yet developed for evaluating the production of a commodity. It has the most comprehensive approach to sustainability developed to date. It encompasses concerns for sustainability beyond on-farm activity to include the production of wines. It is the first partnership to evaluate operations on the basis of personnel practices and how they relate to their neighbors and communities.[13]

As I explained in chapter 6, the network of cooperators in winegrape partnerships extends to the processors (wineries) more than any other crop. The Code workbook provides quantitative, objective criteria for assessing current practices throughout the state. The Wine Institute and CAWG intend the Code to validate the notion of self-regulation. If they can hold their members voluntarily to a higher standard than other agricultural sectors, they expect to avoid regulatory scrutiny.

Commodities Have Varied Capacities to Circulate Knowledge to the Public

Differences in the commodity form of different crops shapes the ability of a particular commodity network to circulate knowledge throughout the agricultural knowledge system to "the public," whether public regulatory officials or the consuming publics. Commodity-specific groups are more vulnerable to negative public opinion about production practices to their specific farming systems—and are thus more motivated to address them—than are general agricultural lobby groups such as the Farm Bureau. Several commodity organizations have used crop-specific and region-specific knowledge about agroecological strategies and practices learned through partnerships in negotiations with public regulatory agencies.[14] From a pragmatic political perspective, commodity organizations are in a stronger position to negotiate with regulatory agencies if they can demonstrate "good faith" on the part of their growers through partnership initiatives. The PMA program in particular provided them opportunities to engage the Department of Pesticide Regulation. Both almond and pear commodity networks have close relationships with their processors, but differences in their economic relations with their processors shape the circulation of knowledge from the farm, scientific community, and growers to the public.

The Almond Board of California and similar groups may in fact be more fearful of negative publicity than actual enforcement of regulations. Several partnership leaders suggest that agricultural commodity organizations worry that another "Alar-type" scare would hurt their industry more than a regulatory action. For example, even though the Sun-Maid raisin partnership achieved substantial reduction in hazardous agrochemicals, Sun-Maid declined to release results demonstrating that for this study. They manifest a strategy of not drawing consumer attention to the use of agrochemicals in agriculture, but should a food safety crisis erupt, they could pull the data out of a drawer to deflect public criticism.

The almond industry has been much more effective communicating with public officials than with consumers about partnership activities. Even though PMA funding has come to an end, the ABC has found it to their advantage to continue the partnership. All the parties in the network brought significant resources that were made more available to other sectors of the network to better achieve common goals. In this regard, the ABC does not have to represent its growers to regulatory agencies as reducing their pesticide use to zero, but only demonstrate a reduction in their use faster than other commodities with similar problematic pesticides, such as prunes. The process of bringing these institutional representatives to negotiate a consensus about the needs of the industry is beneficial *in and of itself* because it helped oriented a group of institutional actors toward common goals. The ability of the ABC to convene diverse interest groups within the industry should not be overlooked. Relationships within many commodity networks are fractured. Some other commodity groups have been unable to agree even on how to go about writing a grant proposal to fund an agricultural partnership, and several have been voted out of existence by their member growers.

The economic structure of almond hulling and marketing firms mirrors that of almond production: a few dominant operations and many smaller ones. This bi-modal economic structure plus the legal constraints on the ABC as a commodity board have so far thwarted eco-label efforts in the almond industry. Blue Diamond processes about one third of California almonds, and is at least six times the size of the next largest processor. None of the other hundred or so processors have more than 5 percent of the market. In 2001, Augie Feder took a leave from the USEPA to try to start a BIOS eco-label. He worked with a marketing consultant to develop certification standards and a marketing strategy for BIOS-grown almonds. Blue Diamond was decidedly cool to the plan. It was uninterested in additional separation and certification hassles, but also feared losing market share to their smaller, more nimble competitors. European demand has kept organic almond prices high, roughly twice the rate of conventional almonds, so smaller hullers who handle organic nuts did not support any effort that could cut into demand for their product. In addition, Blue Diamond represents about 1,300 growers (many of them large) and does not want to antagonize them, nor expose them to any public perception that non-BIOS almonds cause pollution. Blue Diamond growers generally have four of the ten votes on the ABC board of directors.

The ABC, like other state authorized commodity organizations, can only take on activities for which it has official approval. Chris Heintz reports that the ABC has to serve the industry as a whole, and cannot engage in market development for any environmental value-added product. She said: "the industry has been opposed [to a BIOS eco-label] because they think the whole industry is BIOS. Our entire industry is biologically integrated, so why make a big deal about creating a niche?"

The pear processing industry in California is much smaller relative to the almond processing industry. There are only a handful of pear processors, and they all know each other personally because pear production is geographically concentrated in four small regions. (Recall figure 6.1.) Pear growers prefer to sell to the fresh market because it is more profitable, so the processing companies work to maintain good relationships with the growers because they need a safe and consistent flow of food products. The jointly funded Pear Pest Management Research Fund, which supported the Randall Island Project, is an example of this cooperative approach.

Many canned pears are sold under brand names, such as Gerber and Del Monte. Both of these companies are well aware that they profit from the public perception that their brands offer a superior product, and they have expanded their definition of quality to include food products grown with relatively more agroecological production techniques. Gerber is particularly sensitive to food safety issues because its baby food sells for a premium because of the confidence mothers (and their mothers!) have in its brand. Its entire customer base turns over every year, so nourishing consumer confidence has to be integrated into its business plan. In the words of Nick Heather, director of product safety at Gerber: "we live and die by consumer trust."

Gerber spends about $250,000 a year across the United States to fund research and pest monitoring that will reduce the use of hazardous pesticides and increase IPM skills among their regular growers. Gerber has spent roughly $2 million over the past two decades, much of it on pheromone research for tree fruit, in what they think of as a "supplemental extension service" for growers with whom they have a long-standing relationship. Heather justifies this expense in part because Gerber has instituted a program of traceability for food products through its processing run. Heather and Gerber know they cannot draw attention to the fact that pesticides are used in agriculture, so they recently conducted an advertising campaign with images of Michigan family farmers and the slogan: "do you know where your baby food comes from?" Like Sun-Maid, Gerber is doing everything it can to avoid being tarred with the next Alar scare, but does not want to talk about pesticide reduction because it does not want to talk about pesticides at all. Gerber is negotiating with Protected Harvest for certification as well.

Conclusion

Agriculture is an economic activity, and by coupling economic rewards with agroecological knowledge, partnerships accelerate the circulation of science through networks. This does not debase agroecology, but on the contrary, makes it more viable, more powerful, more influential. Latour would argue that these economic incentives are just as much a part of agroecology as is the understanding of biological control or cover crops. Economic advantage is a powerful incentive for social learning. Ohmart and the Lodi partnership have created a full-blown working model that reflects Latour's circulatory system of science. This system is doubly fed by agroecological knowledge linked with economic incentive. They have knotted economic and environmental quality.

   People work better together when they have reason to collaborate. The partnership participants this chapter describes have recognized that it is in their own best interests to work together. What they perceive as the greatest economic threat is not competition from their neighbor growers but competition from growers of the same crops in other countries. Sustained collaboration has helped them overcome production and public representation problems over the past decade, and appears the most likely strategy to help American agriculture survive in the future.

8

Public Mobilization

Green Potatoes

The Great Lakes—the most important source of water for the north-central and the northeastern United States—are quite vulnerable to human degradation. More than 30 million people depend on the Great Lakes for drinking water, recreation, and economic uses. Agricultural non-point-source pollution is a major source of contaminants for the lakes. The US Environmental Protection Agency has devoted significant resources to addressing water-quality issues in this area since its inception. In the 1990s, the World Wildlife Fund held a series of regional stakeholder meetings to see what could be done to prevent pollution, especially chemicals causing endocrine disruption, due to their serious and lasting damage to fish, bird, and animal (including human) life in this area.[1]

In the course of these meetings, World Wildlife Fund staff members learned that many IPM tools existed, at least on the shelf, but that they were not being adopted in large numbers. After the passage of the Food Quality Protection Act, the WWF had a series of conversations with Dean Zuleger, the executive director of the Wisconsin Potato and Vegetable Growers Association. Zuleger was concerned about how this act would affect his 350+ growers. Wisconsin potato production depended heavily on soil fumigants, insecticides, and fungicides, and he felt his industry was vulnerable to catastrophic disruption due to the loss of pesticides under the Food Quality Protection Act. The University of Wisconsin had developed IPM practices, but few association growers were using them. Zuleger and WWF staff brainstormed. Was there an opportunity hidden in this looming crisis?

Wisconsin ranks third in potato production, but far behind Idaho and Washington. Could Wisconsin growers make environmental stewardship an opportunity for adding value to their crop? Could the WWF bring positive attention to such an initiative? The WWF was interested in helping environmentally responsible growers, but was understandably cautious about using conferring legitimacy on any activity associated with pesticides. The WWF was not content only to help the environmentally oriented growers; it also wanted to facilitate pollution prevention by all growers, including those using the most pesticides. The challenge became creating incentives to keep all growers moving along the continuum—so that average growers progressed toward biointensive IPM practices—while continuing to identify new, innovative practices.

In 1996, the Wisconsin Potato and Vegetable Growers Association and the World Wildlife Fund drafted a memorandum of understanding, setting targets for eliminating certain pesticides, but also goals for the entire association of growers. Sarah Lynch from the WWF became the project manager, and the partnership engaged the services of Chuck Benbrook to provide technical support and analysis. He had developed the "IPM continuum" framework. Because Integrated Pest Management is a process or an approach to controlling pests, most growers and agricultural organizations claim to be using it. Informed by agroecological principles, Benbrook distinguishes between low IPM (merely scouting the presence of pests), high IPM (using a variety of preventative tactics), and biointensive or bioIPM (which uses biological strategies whenever possible).[2] The partners agreed to

set ambitious goals and timetables for measurable bioIPM adoption and pesticide use,

promote research and extension emphasizing alternative practices,

create indicators for measuring progress in eliminating the most hazardous pesticides and adopting alternative practices (including "reduced risk" pesticides),

collaborate to develop opportunities to add value to potatoes grown with bioIPM,

and

develop strategies for enhancing biodiversity.[3]

University of Wisconsin extensionists and the association worked with small groups of progressive growers to develop alternatives, and then engaged them in outreach to other growers. The WWF pledged to work with the Wisconsin Potato and Vegetable Growers Association to gain marketplace recognition once the association had achieved certain toxicity reduction goals with 11 "high-risk" pesticides.

The association leaned heavily on its long standing relationship with the University of Wisconsin, and the Potato IPM Team at the Madison campus officially joined the partnership in 1999. The university and the association had begun collaborating on IPM in the early 1980s after aldicarb, a carbamate pesticide, was found in drinking-water wells in rural Wisconsin. These two institutions recognized that nearly everyone claimed to be using IPM, but not all were in fact defining it the same way. When the university and the association wanted to examine detailed, specific information about pesticide use for their partnership, these growers were willing to share data about their farming practices and pesticide use.

The partnership struggled to define a meaningful measure of progress because "pounds of active ingredient" is not an accurate measure of hazard. In addition, while the focus was on 11 high-risk pesticides, the partners needed to find out if growers would compensate by increasing their use of other high-risk pesticides. They concluded there is no one correct way to measure pesticide risk, and that every approach has advantages and disadvantages. In the end they worked with Benbrook to develop an approach for balancing acute mammalian toxicity, chronic toxicity, ecotoxicity, and disruption to bio IPM, and calculated a "toxicity factor" for every pesticide used in Wisconsin potato production.[4] They established a baseline of 1995 use levels from USDA data, and then surveyed association growers.

The 1997 survey showed a 28 percent reduction in toxicity units, and the 1999 survey 37 percent reduction since 1995. Some growers had substituted other high-risk pesticides, but many used new, softer materials instead, even though they are more expensive than older, more hazardous and effective pesticides. As a part of this project, University of Wisconsin researchers discovered that by increasing the distance of potato fields from the previous year's planting to 400 meters reduces damage by the Colorado potato beetle 90 percent, reducing the need for insecticides.[5] Wisconsin potato growers (unlike those in the West) already used a three year rotation with vegetables and grains, meaning that they were able make even greater reductions in insecticide use. This strategy will have an even greater impact as growers are able to coordinate their plantings with their neighbors.

The WWF recognized that it did not want itself to be in the business of certifying growers' practices, so as the Wisconsin potato partnership began to show signs of success, it helped create Protected Harvest. This is an independent certification organization, with the mission of certifying groups of growers who use sustainable agriculture practices. Unlike organic certification, Protected Harvest works with groups of growers who develop stringent, transparent, and quantifiable standards, which Protected Harvest then certifies. The Wisconsin potato partnership was its first initiative, but it is now working with several crops in several states, including the Lodi partnership.

The Wisconsin potato partnership, working with Chuck Benbrook and Jeff Dlott, set standards for "Healthy Grown" potatoes, based on production practices, toxicity impacts, and chain of custody. Protected Harvest started certifying these potatoes in 2002. Until Protected Harvest developed its own identity, Wisconsin "Healthy Grown" carried the internationally recognized the WWF's panda logo. By 2005, about one quarter of the fresh market potato growers of the Wisconsin Potato and Vegetable Growers Association had met the threshold established to be certified, enrolling about 10,000 acres. They were only sold as fresh market potatoes in the Midwest and in the East.

In 2003, US Department of Agriculture Secretary Ann Venneman recognized the partnership for its innovation. The Wisconsin potato growers are very proud of what they have done. They have come to recognize that regardless of how they feel about change in agriculture, it is inevitable. The partnership with the WWF and the university has given them the opportunity to shape the trajectory of change. The simple fact that growers were willing to speak in public about their use of "toxicity units" is a testament to their willingness to address the environmental impacts of conventional farming with courage.

The University of Wisconsin researchers have continued to expand their work for the partnership, funded in part by more than $1 million dollars in grants from a variety of sources. They have broadened their focus from bioIPM to integrated farming systems, and now to the landscape of which farms are a part. Deana Sexson, the bioIPM field coordinator for the university, wrote her dissertation about area-wide pest management in Wisconsin potatoes. Recently the university team began investigating how to best manage non-crop areas for farm pest management and wildlife habitat. The International Crane Foundation is now working with Sexson to determine how Healthy Grown potatoes could include habitat conservation for endangered sandhill cranes, and arrange for Protected Harvest to certify these as ecosystem standards.

In addition, Deana Sexson and her team are now turning their systems approach to soil ecosystems. Since Wisconsin potato growers use a lot of soil fumigants for diseases, the partnership identified this as the next major issue to tackle. Sexson and others are trying to determine how soil improvement strategies could reduce reliance on fumigants, and has set the ambitious goal of helping reduce their use by 50 percent by 2007. She hopes to create a list of management options, such as tillage, nitrogen and other fertilizers, cover cropping systems, deep rooted systems, and other cultural practices.

The Wisconsin potato partnership helped the WWF realize the value of engaging agriculture. As a global conservation organization, they know that simply driving agriculture out of the United States with environmental regulatory pressure will only result in the further loss of wildlands in the developing world, where even fewer regulations are in place. Keeping US agriculture viable while reducing its impacts results in net global environmental progress. The WWF's constructive engagement with the Wisconsin potato partnership has persuaded them that it is possible to have a positive impact on conventional agriculture, especially by creating incentives to eliminate the most environmentally problematic practices. As an organization, the WWF understands that growing food without any pesticides is unrealistic, but that it is possible to create incentives to move a group of willing growers along the bioIPM continuum. Like many other partnerships initiated by non-farming organizations seeking to improve stewardship in agriculture, the WWF has learned the value and power of working with commodity organizations. The WWF has been able to raise the issue of pollution prevention to the national agricultural policy stage.

Lynch and the WWF are particularly pleased that the Wisconsin potato partnership has been able to move beyond merely pesticide reduction.

They are enthusiastic about the possibility of including ecosystem management standards, and are thrilled the partnership is willing to tackle this, just as they are the soil fumigation issue. According to Lynch, this partnership spawned greater awareness of the possibility of working with agriculture on the part of the WWF. Now agriculture is a much more of a focus in the organization.

As part of her work with WWF, Lynch recently convened a partnership in Florida with cattle ranchers, who rear calves in the winter and then send them west to grow into adult cows. Much of the state of Florida was a wetland, and a century of draining and agriculture production has had terrible environmental impacts on Lake Okeechobee. Phosphorus runoff has created toxic algal blooms across most of the lake, which threatens the multi-billion-dollar Everglades restoration plan. The WWF has assembled a partnership of private cattle ranchers, the Florida Department of Agriculture, the US Department of Agriculture, the National Resource Conservation Service, and the local water-management district. The WWF has raised $2 million for a 3-year project to develop, field test, and document a land and water stewardship partnership, with an eye to scaling it up across the state. Lynch has worked for several years to scope out the economics and build up the level of trust to the point where partners can work together.

The WWF does not use its panda logo casually or often. Every prospective use is subject to vigorous internal debate: what is the most powerful way to use and protect it? Lending it to any product has risks, and these need to be assessed against the potential environmental gain achieved by informing consumer choice in the marketplace. The WWF helped launch Protected Harvest to further this work, and to create a strategy for conferring legitimacy on agroecological initiatives without having to use the WWF's panda logo. Some within the organization see the logo's association with an independent third party certification effort as a very successful, programmatic use of the logo, while others are still uneasy with this idea. The partnership with Wisconsin potatoes certified by Protected Harvest was a 3-year pilot use of the logo that ended in 2006, and it is not currently being used elsewhere, although it may in the future.

The Healthy Grown potatoes partnership is a remarkable working model of agroecology. Growers, their organization, scientists, and an environmental non-governmental organization were exceptionally entrepreneurial in its development. They assembled all the necessary pieces to this model, and circulated knowledge widely. Protected Harvest works with agriculture to verify the progress that groups of growers make toward environmental goals and reduced impacts from pesticides. Protected Harvest is poised to forge a vital link between environmentally responsible growers and the consuming public. Protected Harvest is unusual among third party certification agencies because it works to certify local groups of growers. This means that pre-existing cooperative agroecological initiatives, such the Healthy Grown partnership or Lodi Woodbridge Winegrape Commission, are able to capture further benefits from their historical social networking. The USEPA awarded Protected Harvest a $425,000 grant in 2005 to develop standards for dairy, almonds, tomatoes, and stone fruit in California.

Protected Harvest provides a critical link between agroecological initiatives and the public, yet it is unfortunate that this kind of organization is needed. Much of the consuming public has lost trust in the American industrial food system because the agricultural industry and policy makers have for too long failed to address its negative environmental and health impacts. For too long, the United States has only measured agricultural benefits in terms of cheap, voluminous food. Protected Harvest provides economic rewards for groups of entrepreneurial growers, and that is good. It helps environmentally conscious consumers to have confidence in the health and stewardship of some food products, which is also good. For a genuinely healthier relationship between food, farming, and society in the United States, however, we need a broad, vigorous debate about how policies can foster the kind of agriculture we want, and the science needed to support that.

The Logic or Illogic of Agroecology in Advanced Capitalist Countries

Agricultural science is like the tip of the spear—it indicates the trajectory of progress, for better or worse. The science of agroecology requires careful observation of nature as an integrated system that varies across time and space, which requires specialized knowledge and labor. The environmental harm caused by agriculture is rooted in simple technologies that have serious impacts when deployed across a wide scale.

Agroecology helps us recognize that unintended environmental consequences of these technologies are, from a systems perspective, entirely predictable. The monocultures which dominate American agriculture are economically rational but ecologically irrational. The economic efficiencies of labor saving technologies such as tractors and agrochemicals thwart the polycropping and bio-diversification strategies of agroecology more common in the developing world. Any effort to deploy agroecology as an environmental problem-solving strategy in the United States must confront the economic rationality of monoculture. Genuine progress can be made, however, using the five strategies described in chapter 5, even within the framework of economic monoculture. The agroecological initiatives this book describes have occurred despite the counter-incentives of the dominant institutions in American agriculture.

Agroecology cannot be transferred as a technological package. It can only be facilitated by social learning. It is inherently more knowledge intensive than conventional approaches to agricultural production. As chapter 5 demonstrated, monitoring data must feed integrated farm management decisions so that growers can realized the benefits from managing systems interactions. The agroecological partnerships described in this book manifest a critique of conventional "Transfer of Technology" extension practice and the simplistic and hegemonic assumptions it contains. Agroecology in the United States relies upon the expertise of scientists, but agroecological partnerships have had to work very hard to secure their participation in networks. The dominant incentives in agricultural science do not reward extensionists or research scientists for contributing to applied projects such as partnerships.

The Significance of the Agroecological Partnership Model

The narratives and case study in this book demonstrate that voluntary change in agriculture is possible—indeed, that it may be the only viable strategy for improving environmental stewardship in US agriculture. Environmental NGOs and environmental protection agencies have concluded that conventional regulatory strategies alone are unlikely to make authentic progress. Restricting the use of toxic or hazardous pesticides is one necessary tactic of a coordinated strategy, but this by itself will not spur the kind of changes necessary. Agroecological initiatives require social learning, which regulation alone cannot achieve.

Few agroecological partnerships have been as successful as the Lodi winegrape partnership or the Wisconsin Health Grown potatoes project. We should not conclude that every agroecological initiative could be as successful as they are. The genuine significance of these partnerships lies in the way these participants joined together in a social learning effort engaging all four Latourian loops, and linked scientific, economic and social knowledge to strengthen the ties that bind them together. Latour's model calls our attention to the wide range of social activities necessary to successfully deploy scientific knowledge. Other groups of growers, in other commodities (especially those federally subsidized crops), may have had access to more sophisticated scientific knowledge about their crop production, but they have not integrated agroecological knowledge into their social networks.

Local or regional groups of growers seem to be better positioned to facilitate the kind of social learning required by agroecology. Small groups of neighboring growers can develop new practices, but they often lack the influence to achieve broader, systemic change. Large, national commodity groups generally appear uninterested in agroecological initiatives. This may be because they are deeply invested in farming the existing federal crop subsidies, or because very large corporate interests dominate the organizations, or because food processing groups such as the national association of corn millers merely want to continue business as usual. Medium-size commodity groups of growers have sufficient resources to organize new initiatives, yet remain responsive to their member growers.

A few NGOs, including the World Wildlife Fund, have engaged in partnerships because they want to demonstrate that improved environmental stewardship in agriculture is possible. The working model they helped shape is commendable, but the WWF does not have the resources to address the full scale of the agro-environmental problems this country faces. The US Environmental Protection Agency, the California Department of Pesticide Regulation, and similar agencies have made a positive contribution to agroecological initiatives, but the impact of their participation might have been greatest within their agencies. Elected officials have not passed necessary legislation to authorize environmental regulatory agencies to contribute to these initiatives. Creative agency staff have had to stretch existing programs to provide the kind of assistance needed. Protected Harvest is more nimble and entrepreneurial than regulatory agencies with their limited legal mandates, but one cannot avoid the inference that it exists as an organization because public officials have not developed responsible agro-environmental legislation.

Latour's circulatory system of science model is particularly helpful for interpreting how diverse actors reshape an agricultural knowledge system to meet their needs. It has helped us interpret the dialectical relationship between these actors and agroecological knowledge. The growers described in these narratives believed that another way of farming was possible, and set out to discover how to improve their practices. They learned as most farmers have learned throughout time: by trial and error. This approach and the scientific method yield different kinds of knowledge. Trial and error cannot establish causal relationships like the scientific method, but these case studies show how it can foster innovative thinking.

For most of the twentieth century, most American agricultural scientists have fallen into the trap of reductionistic thinking. Agroecology is the antidote. Genuine environmental progress requires a systems perspective. Latour's circulatory system of science does a superior job of explaining how science really works, especially during these periods of transition. STS perspectives help us to see how science is never "pure content," but rather woven into social networks, tied to existing motivations.

Agroecological partnerships should not be thought of as permanent features. They most often arise during a period of transition or uncertainty. The additional investment in social networking can only be justified so long as participants are able to realize benefits from their interactions. After several years of close collaboration, partnership participants experience diminishing returns on the investment of their time. After this flush of new social learning, growers and their consultants can approach their farming systems in new ways. Here partnerships again confront the existing structures of economic incentives.

Could partnerships operate successfully on a broader geographic scale? A nascent partnership in the Mississippi River watershed is poised to test this question. The Green Lands, Blue Waters project is an attempt to develop a new generation of farming systems in this region, integrating more perennial crops and continuous living cover into the agricultural landscape.[6] "Continuous living cover" indicates agricultural cropping and livestock production systems designed around perennial crops, such as trees, shrubs, and grasses. This project integrates scientific, social, and economic research and practices that will improve water quality, increase grower profitability, improve wildlife habitat, reduce flooding risks, and enhance rural community vitality. This is the most ambitious agroecological partnership yet conceived, yet it is precisely the kind of initiative our country needs. More than two-thirds of the land in the Mississippi River watershed is farmed, most of it in annual crops, leaving soils and nutrients and agrochemicals highly vulnerable to erosion, which in turn creates the conditions for hypoxia in the Gulf of Mexico. The Green Lands, Blue Waters project seeks, in the words of Wes Jackson and Jerry Glover, a perennial solution to farming's annual problem, or what Jackson calls "the problem of agriculture."[7] This project has enrolled five land-grant universities in the basin. The Leopold Center for Sustainable Agriculture has signed on for Iowa State University, however, and it is not clear how much its parent university is invested. Project organizers anticipate a $105 million budget over ten years, not including farm incentive payments. The ability of projects like this to effect change appears largely at the mercy of American farm policy.

Agroecology: Scientific, Economic, and Policy Challenges

This book has used Latour's model to interpret the development of alternative agricultural science, and we have seen that progressing toward integrated agro-environmental goals depends upon organizing alternative social relations. What policies could better support the social relationships to put agroecology into action on a broader scale? The land-grant university and the federal crop subsidy systems are the two dominant institutions in American agriculture, and agroecological partnerships have developed in spite of their counter-incentives. The USEPA has never been authorized to manage the health of ecosystems. To realize the benefits of agroecology will require more than its public representation as an alternative, but rather reforming and re-directing these institutions, but this will not happen without mobilizing the public to support such efforts. I will close by offering a few suggestions for policy reform.

First, the public should demand publicly funded land-grant universities provide the kind of science to support the agro-environmental leadership that society needs. Latour's model suggests that scientific expertise is critical, but that scientific experts have to be recruited to participate in but not dominate agroecological initiatives. Land-grant universities tend not to reward applied environmental problem solving. They have been the subject of continuous criticism since Hightower for failing to protect the common good. In general, LGUs have drifted away from serving the needs of farmers and rural communities because public officials have failed to support these kinds of programs with funds, and because private industries have stepped up to capture their attention. The criticisms initiated by Carson and Hightower raised public questions about industrial agriculture, but the inadequacy of response to these questions has contributed substantially to a loss of public trust in agriculture and in LGUs.

That only a dozen land-grant universities have sustainable agriculture centers is absurd in view of the enormous pressures American agriculture faces on every side, and serves as further indictment of LGU administrators' inability to adequately recognize growers and rural communities as their primary clients. (Recall figure 2.2 and its "new model" of LGU research.) Every LGU should have a sustainable agriculture program capable of organizing agroecological initiatives, and facilitating the participation of scientific experts. The public—through their elected officials—should demand this. In reality, only some LGUs will consider these kinds of initiatives. The LGUs in some states have demonstrated a willingness to take up this challenge, expressed through alternative agriculture centers in Wisconsin, Iowa, and Washington. They continually have to justify and defend their budgets, but they are models for responsiveness, at least relative to the more obdurate example of California.

The leadership of the University of California tried to abolish the Sustainable Agriculture Research and Education Program in September 2003, and this effort can be traced to decisions at the highest levels of its administration. Only the outrage of sustainable agriculture organizations like the Community Alliance with Family Farmers and the

California Sustainable Agriculture Working Group prevented this from coming to pass, but UC will try again. Agroecological partnerships brought less than a million dollars annually in extramural funds between 1994 and 2004. Relative to private sector grants for agricultural biotechnology research, alternative agriculture just does not pay enough. UC's active efforts to suppress dedicated IPM and sustainable agriculture programs seem all the more petty in this context of their modest costs.

The University of California's lack of interest in agroecological initiatives is ironic in at least three ways. First, UC now operationally defines its primary clients as input suppliers, not growers, taking Charles Woodworth's recommendation of a century ago to an extreme. Second, UC denies a role for public interest NGOs in setting its agenda, even though it is substantially supported by public funds. Third, it solicits private industry funds—and their strings—that compromise its public interest charter while actively working to squelch legislative public funds that would advance its public mission. UC leaders often behave as though they were beyond the reach of legislative input and oversight. Given the history of public criticism of LGU research impacts, the unwillingness of UC leadership to support agroecological partnerships is all the more scandalous.

Other major farming states, such as Florida and Texas, do not even have alternative agriculture programs, and the public extension services in many other states are being dismantled due to budgetary pressures and the inability of public officials to recognize the public good provided by them. Agriculture varies tremendously by crop and region, and so do LGUs. Agroecological initiatives will have to take these differences into account.

For the immediate future, agroecological initiatives will be organized and animated by leading growers and their groups. Partnerships are thriving among winegrape organizations, and some practices promoted by pear and almond partnerships have apparently been adopted widely by growers. But these represent privatized, or at least semi-privatized initiatives, and the difficulties they report in securing the work of public scientists defies common sense. Agroecology will never break out of its jacket of "alternative agriculture" without sustained efforts to engage agricultural science institutions. Non-governmental organizations will have to find new ways to mobilize the public to pressure these public institutions to foster agricultural science in the public interest, and providing resources to these kinds of organizations is a critical role for private, philanthropic foundations.

Second, the US Congress should reform the out-dated and misguided federal crop subsidy system, and replace it with public support for land stewardship. Chapter 4 of the 1989 *Alternative Agriculture* report by the National Research Council was titled "Economic Evaluation of Alternative Farming Systems." It begins:

Interest in alternative farming systems is often motivated by a desire to reduce health and environmental hazards and a commitment to natural resource stewardship. But the most important criterion for many farmers considering a change in farming practices is likely the economic outcome. Wide adoption of alternative farming methods requires that they be at least as profitable as conventional methods or have significant non-monetary advantages, such as preservation of rapidly deteriorating soil or water resources.

Many sustainable agriculture programs based at LGUs are now investing less in alternative practices and more in alternative markets. The concentration in the industrial agrofood system continues to squeeze small and medium-size farmers out of markets, and concern for one's economic survival understandably trumps improving environmental stewardship. One participant in the Practical Farmers of Iowa said: "I don't want to be the last farmer in my county doing a nutrient budget." Developing agroecological practices makes no sense to a farmer who cannot pay the bills. For more agroecological strategies and practices to be put into action will require dismantling the existing federal system of subsidizing selected crops and replacing it with an incentives system that encourages stewardship. A tiny fraction of California agriculture receives federal subsidies, and this appears to have enabled the extraordinary scale of agroecological innovation here. At present, American farm policy provides incentives for wasteful and harmful agricultural practices, and penalizes farmers for developing alternative crops, practices, and markets. Agroecology will remain a marginal practice so long as it is a personal virtue, only of interest to a few leading growers.

The 1996 farm bill, known as the "Freedom to Farm" bill, marked the first effort to reform the New Deal commodity payment programs by allowing farmers to rotate crops without losing crop support. It also included the Fund for Rural America, which was conceived as a strategy for shifting some of the support for commodities to more diverse, applied socio-environmental problem solving initiatives. This fund was an excellent example of how to support alternatives to conventional agriculture. It was authorized in the farm bill, but its funding was later diverted to other programs. Public interest groups rejoiced at its inclusion in the farm bill, but were unable to defend it in the appropriations process. This kind of policy would have substantively supported agroecological initiatives, and still could, if it were properly funded.[8]

Since the publication of *Alternative Agriculture*, many scholars have begun to describe agriculture as multi-functional, meaning that its social benefits are more than merely commodity output. For all their stated opposition to ending agricultural subsidies, elected officials in the European Union have quietly been funding policy makers to craft alternatives to their agricultural production subsidies. The industrial countries need agricultural policies that recognize the multiple social benefits created by farming, such as ecosystem services, known as multi-functionality. Ultimately, agriculture will have to do more than merely achieve arbitrary water quality goals: it will have to support resilient ecosystems. Farming is about more than food, whether modern growers like that fact or not.

The farm-belt states in the Mississippi River basin receive the lion's share of federal subsidies and appear to have the most to lose from restructuring federal subsidies. The hypoxic zone in the Gulf of Mexico is the indirect but inevitable result of the federal crop subsidy system, and cannot be substantively addressed without reconfiguring this system. A farm policy obsessed with ever greater output commodity production cannot, by definition be sustainable. The agricultural input and processing industries have benefited from crop prices distorted by these subsidies, and have effectively curried elected officials to keep them in place. Advocates tried to reform these during the 2002 Farm Bill debate, but were unsuccessful. President George W. Bush has pledged to address this crop subsidy system as a bargaining chip for the World Trade Organization negotiations, but his ability to deliver on this is uncertain. Some policy makers have discussed the creation of a "green box" in the next farm bill, a strategy for providing public funds to support for agriculture for environmental stewardship, instead of merely production. This would be a mechanism for bringing US farm policy into compliance with World Trade Organization guidelines. To be successful, a green box would have to have some mechanism for accountability, some means to foster public trust that this policy is achieving its stated goals. Protected Harvest is a working model of how to verify this progress with scientific data.

Third, Congress should create a comprehensive agro-environmental policy and fund agroecological initiatives. Congress has never provided proper direction or authorization for the USEPA to help agriculture meet its pollution-prevention goals. This is a major impediment for the agency, and for environmental protection in general. The USEPA should be given authority to foster healthy, resilient ecosystems, through its own initiatives and in partnering with the Department of Agriculture and other agencies.

American agriculture is worthy of public support, but only if it provides the kinds of socio-ecological benefits required of a good neighbor industry. Agriculture in all US coastal states is straining under land use pressures. Environmental leaders in these states in particular should recognize the importance of agriculture in providing ecological services, and devise collaborative efforts to its preservation. Mark Chandler, the Executive Director of the Lodi Woodbridge Winegrape Commission, asserts that the best defense against suburban sprawl onto prime farm land is to ensure the economic viability of agriculture. Agroecological initiatives, held accountable through programs like Protected Harvest, can help restore trust in agriculture, and contribute to mobilizing the public to demand a different kind of agriculture.

For decades the public has largely accepted the "family farm" myths as a representation of contemporary American agriculture when it is in fact a large scale industrial activity. Agriculture has as much if not greater impacts than any other economic enterprise on human health and the environment. Agroecology holds out the scientific promise and implicit ethos to transform American agriculture. Mobilizing the public to demand this of our elected officials will be required to further put agroecology into action, and achieve Rachel Carson's dream of a healthy relationship between society and nature.

Notes

Introduction

1. See FitzSimmons and Goodman 1998.

2. Throughout this book, "pesticide" refers to insecticides, herbicides, fungicides, and rodenticides. For national trends, see Aspelin 2000, p. 4.9. Much of this increase is due to the reliance on herbicides in the midwestern farm belt.

3. In 1991, Congress funded the US Geological Survey to conduct a major survey of US water quality, the National Water-Quality Assessment Program (NAWQA, pronounced "naw-qua"). NAWQA's research focuses on more than fifty major river basins. For a national overview, see US Geological Survey 1999.

4. See Carpenter et al. 1998 and Howarth et al. 2000.

5. See Harrison 2004.

6. See Colborn et al. 1996 and Hayes 2005.

7. Carson 1962, p. 263.

8. National Research Council 1993, but see discussion below.

9. For an analysis of social learning for environmental protection, see Woodhill and Röling 1998.

10. Röling and Wagemakers (1998) and Hassanein (1999) investigate social learning processes and networks in sustainable agriculture, but they do not explicitly use agroecology as a framework for investigating extension processes.

11. My understanding of the centrality of knowledge questions in sustainable agriculture draws on Jack Kloppenburg's (1988) groundbreaking work *First the Seed: The Political Economy of Plant Biotechnology* and on his 1991 article on the de-/re-construction of agricultural science for alternative agriculture. Neva Hassanein (1999), his doctoral student, substantially shaped my understanding of the critical role of knowledge and power relations in social networks for sustainable agriculture.

12. Warner 2004.

Chapter 1

1. Named individuals were the most important sources of information for these three vignettes. For a complete discussion of the California case studies, see the methodological appendix in my dissertation (Warner 2004). California's pest-control advisors are described in chapter 4.

2. Information on codling moth resistance from Dunley and Welter 2000.

3. This according to Benbrook (1996), a leading scientific critic of environmentally harmful pesticide use.

4. The former two are compatible with organic production. Mites are arthropods closely related to spiders that are not members of the class insecta. Miticides are pesticides used to control them.

5. As documented by the PUR database. For details of this database, see below. For information on organophosphate reductions in the California pear industry, see figure 3.6.

6. See Epstein et al. 2000, 2001; Zhang et al. 2004; Elliott et al. 2004. I analyze and interpret the data at the end of chapter 3.

7. Crush districts were established by the state legislature in the 1880s. Some correspond to county boundaries, but those in the San Joaquin Valley do not. The Lodi district includes southern Sacramento and northern San Joaquin Counties. For more information on local and state wide agricultural districts and organizations, see below.

8. Ledbetter is quoted in National Resources Defense Council 1998.

9. From Culver 1993.

10. For a description of the early Lodi partnership activities, see Culver 1993.

11. For detailed analyses of the immediate political impacts of Carson's work, see Hynes 1989 and Dunlap 1981.

12. For analysis of US environmental policy initiatives shaping agriculture, see Andrews 1999, pp. 170–172, 236–237, and 304–308. The federal EPA is identified hereafter as the USEPA to distinguish it from the California EPA.

13. For a detailed history of how US policy has handled pesticides, see Bosso 1987.

14. On the Clean Water Act, see Andrews 1999, pp. 236–237 and p. 417 n. 17.

15. As Andrews (1999) points out, once point sources (discharge pipes) were effectively regulated, agriculture became a leading source of water pollution, contributing as much as one-third, even though it represents only about one-fourth of US land use. Aspelin (2000, pp. 3–21) estimates total US land area to be 1.9 billion acres, and crop land to be 460 million acres, ranking third behind grassland pasture and range (589 million acres) and forest (559 million acres). His primary data sources were the US Department of Agriculture (USDA) and the US Environmental Protection Agency.

16. For a discussion of how this myth has guided policy, see Browne et al. 1992.

For a broader discussion of US agricultural policy and its dynamics, see Browne 1988, 1995, 1998. The political reluctance to address environmental problems in agriculture continues despite the fact that agriculture is among the most heavily regulated economic activities in the US.

17. See Andrews 1999 and Hendrickson and Heffernan 2002.

18. Regulating agriculture is seen by many as impinging on land-use decision making, a process that has been jealously guarded by local governments at their domain. Land-use planning is known as the "L-word" at the USEPA because of the political conflicts that surround any federal effort to address it (Rosenbaum 1994).

19. For an analysis of these challenges, see Rosenbaum 1994.

20. See National Research Council 1993 and US Geological Survey 1999.

21. National Research Council 2000, p. 2. On hypoxia outbreaks in US coastal waters, see Pew Oceans Commission 2003, p. 22.

22. Although this organism and its threat to human health is poorly understood, it appears that exposure to waters where toxic forms of *Pfiesteria* are active may cause memory loss, confusion, and a variety of other symptoms including respiratory, skin, and gastro-intestinal problems.

23. US Geological Survey 1999, pp. 26–37.

24. US Geological Survey 1998b, p. 22.

25. Source: Aspelin 2000, appendix. These crude figures do not reflect the changes in type of pesticides (insecticides, herbicides, fungicides, etc.) over time. Aspelin estimates agricultural pesticide expenditures from 1961 to 1997 to have risen from $1.5 billion to $4.2 billion (in constant dollars), a rise from 2 percent to 4.5 percent of total US agricultural expenses. Benbrook (1996) drew from USDA agrochemical use surveys and estimated total pesticide use to be about 600 million pounds in the US in 1992. For more on California pesticide use, see chapter 3 below.

26. Aspelin (Aspelin 2000; Aspelin and Grube 1999) estimates US pesticide use at between 800 million and 1 billion pounds. California's proportion of insecticides is much greater than 25 percent because more than half of all US pesticides are now herbicides, intensively used in the farm belt.

27. See chapter 2 of Wright 1990.

28. On the environmental contamination of California, see the following NAWQA reports: US Geological Survey 1998a) on the San Joaquin Valley, US Geological Survey 2000) on the Sacramento Valley. The USGS found the levels of organochlorine pesticides, PCBs, and nutrients in the San Joaquin River to be among the highest in the nation. This river flows through most of state's top agricultural production counties.

29. For more on the environmental impacts of diazanon, see chapter 3.

30. See Andrews 1999, pp. 262–266.

31. See Gottlieb 2001.

32. This split occurred during the early decades of the twentieth century for reasons mostly related to the development of academic disciplines, but since the late 1980s this artificial barrier has begun to erode, at least in some university departments (Elliott and Cole 1989; Gliessman 1998).

33. For a discussion of conservation biology as a value-laden science, see Meffe and Carroll 1994.

34. Altieri 1989.

35. Altieri 2002.

36. See Gliessman 1998 and Altieri 1989.

37. Altieri 1998.

38. Jules Pretty (1998) counted more than 80 different definitions of sustainable development subsequent to the 1987 Brundtland Commission report introducing the term, and he argues that it should never be identified exclusively with a technology or policy (which it frequently has), but rather an overarching set of goals. Hassanein (1999, pp. 2–6) describes the difficulty of using the term "sustainable agriculture" in the US.

39. Allen et al. 1991, p. 37. See also Allen 1993, 2004.

40. Many sustainable agriculture initiatives in the industrial world leave unaddressed questions about social equity, especially regarding farmworkers. Much debate of the term "sustainability" turns on whether a practice or initiative can promote ecological sustainability (resource protection) without simultaneously addressing economic and social relations. For example, Patricia Allen critiques "sustainable agriculture" because it has largely been defined as "a natural/technical process, rather than a social one; a relation between people and nature versus a relations between people *and* people and nature" (Allen 1993, p. 5). Some argue that any initiative must simultaneously address all three, but I leave that particular issue for further debate elsewhere.

41. For a discussion of this discursive framing, see Warner in review (a).

42. On the paradoxes associated with certification, see Guthman 2004a and Guthman 1998. On the origins and ideals of the organic farming movement, see Vox 2000.

43. See Guthman 2004b and Guthman 2000.

44. This field is vast. For an introduction, see Latour 1987, 1999, 1999.

45. To broaden STS's analytical framework, Michel Callon worked with Latour and John Law to develop actor-network theory (ANT). At its core, ANT posits that science and scientists cannot be understood in isolation from their material and cultural worlds. Callon (1986, 1989) developed the initial concepts and language of ANT. For reviews of ANT literature, see Latour 1999 and Whatmore 1999. My work does not deploy ANT, but rather a more practical understanding of socio-ecological networks.

46. Latour 1999, p. 99. Latour explains this model on pp. 98–108.

47. FitzSimmons (2003) applied his theory to ecology and to the Ecological Society of America's Sustainable Biosphere Initiative.

Chapter 2

1. The terms "grass radicals" and "dairy heresy" were coined by Joel McNair (1992a,b). The profile of the Wisconsin intentional rotational graziers and their networks draws primarily from Hassanein 1999. Supplemental information used here is from Cris Carusi and Jennifer Taylor of CIAS and from Hassanein and Kloppenburg 1995.

2. Schlosser 2001, p. 50.

3. Cochrane 1979.

4. Hassanein and Kloppenburg 1995, pp. 726–727 (original citations omitted).

5. Quoted on p. 78 of Hassanein 1999.

6. Cited on p. 77 of Hassanein 1999.

7. For history and analysis, see Rosenberg 1961 and Scott 1970.

8. The information about the Thompsons' farm is from National Research Council 1989 and from Bell 2004.

9. David Lighthall wrote his dissertation on Iowa's ridge-till agriculture and the Practical Farmers of Iowa. He published several articles on PFI farmers' geography, sociology, and strategies (Lighthall 1995; Lighthall 1996; Lighthall and Roberts 1995). Lighthall found substantial progress toward lower input production of corn and soy by ridge-till farmers, but he also discovered that larger (non-PFI) operators viewed ridge-till farming as highly risky. Farmers with smaller farms were able to monitor field conditions and deploy agroecological practices, while larger farmers relied on chemical technologies. Ridge tillage is one of several conservation tillage strategies; see chapter 4.

10. The information about PFI is from Bell 2004, and is supplemented by an interview with Rich Pirog of the Leopold Center.

11. For more on the origins, development, and challenges of the Leopold Center, see Carroll 2005.

12. National Research Council 1989.

13. This history is related in CIAS 1998. For more on the organization and activities of CIAS, see Stevenson et al. 1994. As of this writing, CIAS hosts 16 faculty and staff members engaged in all phases of sustainable-agriculture research, education, and outreach. (Many of these positions are part-time.) See also Carroll 2005.

14. Carroll 2005, p. 59.

15. On the legislation and development of the LGU system, see National Research Council 1995, 1996.

16. Kloppenburg 1988. The adoption/diffusion literature started with Ryan and Gross 1943.

17. On the history of the tomato harvester, see Friedland and Barton 1975 and Friedland 1980.

18. Hightower 1973. On the impact of this book, see Buttel 2005.

19. Buttel 2005, p. 2.

20. Ibid., pp. 5–6.

21. Biological control is "the action of parasites, predators, and pathogens in maintaining another organism's density at a lower average than would occur in their absence." These three ecological actors are collectively known as "natural enemies." The definition is DeBach's (1964), taken from van den Bosch et al. 1982. This definition describes biological control as a natural phenomenon, as a field of study, and as a pest-control strategy.

22. On the original biocontrol campaign and its effect on growers' imaginations, see Sawyer 1996 and Stoll 1995, 1998.

23. Smith's development of the biological control concept and institutions is described in Sawyer 1996. That UC had created an experiment station dedicated to scientific research on one crop demonstrates the power of the citrus industry during this period.

24. This is a central thesis of Palladino 1996.

25. Perkins (1982), Sawyer (1996), Palladino (1996), and Stoll (1998) describe the development of entomological ideas in their historical and social contexts, and the development trajectory of entomology as an academic discipline. Their work substantially shapes my own understanding of relationships between nature, insects, society, scientists, and technology in this section.

26. Stern et al. 1959. Carson lauded biological control, but IPM was not widely known by the time she wrote *Silent Spring*.

27. These principles are adapted from Perkins 1982.

28. Perkins (1982, pp. 76–81) makes reference to the field-based research that led to the formulation of IPM. The case studies in Flint and van den Bosch 1981 also suggest these kinds of local regional collaborations.

29. This history is recounted in Lyons 2004, which is supplemented by California Environmental Protection Agency 2001. For summaries of the UC IPM program's accomplishments and analyses of their funded research projects, see *California Agriculture* 44, no. 5 (September 1990) and 54, no. 6 (November 2000).

30. This term is from Rosenberg 1961.

31. Woodworth 1912, p. 358.

32. On the history of Woodworth and the development of insecticides and their institutions, see Stoll 1995, 1998; Sawyer 1996.

33. Van den Bosch 1978.

34. On the development of the tomato harvester, see Friedland and Barton 1975. On the lawsuit, see Scheuring et al. 1995 and Friedland 1991.

35. Buttel et al. (1993) describe a watershed meeting, sponsored by the Rockefeller Foundation, that launched the "Science for Agriculture" report (Rockefeller Foundation 1982), which spurred the LGUs toward greater scientism. For a UC perspective on this re-orientation, see Learn et al. 1987. For retrospective analyses, see Buttel 2001 and Buttel 2005.

36. The following analysis is from Buttel 2001.

37. Lacy 2000.

38. Röling 1988.

39. Peters (1996) describes the historical and contemporary tensions in extension programs very well. For an older account, see McConnell 1959. There has been surprisingly little work done on the sociology of agricultural extension in the US, and little of what has been done adopts a critical perspective.

40. Fiske (1979) and Scheuring et al. (1995) described the "Farmers' Institutes" in California at the end of the nineteenth century and the beginning of the twentieth.

41. Hassanein 1999, p. 13.

42. See Buttel et al. 1990.

43. From p. 113 of the report, quoted on p. 25 of Peters 1996.

44. This is reflected in a Memorandum of Understanding, sent out by the USDA to the LGUs, that required 75 percent of the national funds to be used for Knapp-type demonstration work (Peters 1996).

45. Peters 1996, p. 22.

46. McDowell 2001.

47. See Warner 2006.

48. For more on consultant-led partnerships, see Glades Crop Care 2003 and chapter 7 of NRC 1989.

49. See Sharp et al. 2000.

50. See Stevenson et al. 1994.

51. Identified in appendix 1 to Hassanein 1999.

52. On the Vermont, Maine, Iowa, and Wisconsin sustainable agriculture programs, see Carroll 2005.

53. On the origins of SAREP, see Scheuring et al. 1995, pp. 217–219.

54. The Council for Agriculture Science and Technology (1990) assembled dozens of reviews by LGU scientists. They reflect the diversity of responses within this scientific community. Kloppenburg (1991) read *Alternative Agriculture* and argued that a different approach to scientific knowledge generation was necessary to achieve the report's goals.

55. National Research Council 1993. Integrated approaches to agricultural production and environmental protection became popular in UK and in Europe at roughly the same time. See El Titi 1992, Morris 1999, and Morris and Winter 2000.

56. National Research Council 1993, pp. 107–108.

Chapter 3

1. Rick Reed, Bob Bugg, Lonnie Hendricks, Glenn Anderson, and Augie Feder were the BIOS pioneers. Hendricks began his study of the Anderson brothers' orchards in 1988, and Bugg and Reed brought their ideas about an extension project to them in 1992. Sources for the almond BIOS case study are presented fully in the methodological appendix of Warner 2004. Pence 1998 substantially informs this case study.

2. The history of these organizations is recounted in Campbell 2001 and in CAFF 2004. After the tomato harvester lawsuit, the California Agrarian Action Project went on to advocate for more environmental laws to regulate pesticides during the early 1980s. In renamed itself the California Action Network in 1985, and then joined with California Association of Family Farmers in 1993 to become Community Alliance with Family Farmers.

3. The California Action Network, the California Association of Family Farmers, the Committee for Sustainable Agriculture, and the California Clean Growers Association came together to form Farmers for Alternative Agricultural Research. The 1990 "Big Green" environmental ballot initiative and its private industry-backed competitor each designated $30 million for alternative agricultural research, but neither of them was passed. Anderson, Buxman, and Masumoto were members of the California Action Network or the California Association of Family Farmers.

4. In 1991, UC Berkeley Entomology doctoral student Jeff Dlott had approached CCGA about conducing his PhD field work in their orchards. Dlott wanted to study insect ecology that could lead to more agroecological practices, but also to investigate the contribution growers could make to agroecological research. For information on the CCGA partnership, see Dlott, Altieri, and Masumoto 1994. CCGA member Mas Masumoto (1995) wrote *Epitaph for a Peach* (illustrated by fellow member and artist Paul Buxman), which describes Dlott's research from the grower's point of view. Both growers had participated in Farmers for Alternative Agricultural Research.

5. On the history of the problem of diazinon, see USEPA 2000a,b. A cursory glance at the list of California's Central Valley waterways failing to meet Clean Water Act water quality goals (the "303d list" named after that provision in the Clean Water Act) reveals diazanon to be a chief contaminant (Central Valley Regional Water Quality Control Board 2002).

6. Initial water quality impacts were reported in Foe 1987, These findings are detailed further in Kuivila and Foe 1995. The report on fog (Glotfelty, Seiber, and Liljedahl 1987) came during the initial era of concern about acid rain.

7. The knowledge controversies associated with BIOS are described in great detail in Pence 1998 and in Pence and Grieshop 2001.

8. Source: Dlott et al. 1996.

9. Private foundation funders included the Charles Stewart Mott Foundation, the Pew Charitable Trust, and the Foundation for Deep Ecology. The National

Resources Conservation Service and Resource Conservation Districts also provided some funding.

10. I count all five counties of concurrent almond BIOS activity as one partnership. The two walnut BIOS partnerships were separated temporally (1994–1998 and 2000–2002) and spatially (Yolo/Solano and Stanislaus Counties).

11. Source of data: Swezey and Broome 2000, p. 30, table 1.

12. For details of the BIFS bill, see Dlott et al. 1996.

13. The 1998 BIFS request for proposals tried to communicate an alternative approach to social relations in agricultural extension. It read: "The projects should use an extension approach that involves public-private cooperation; this approach is often called a 'farmer-to-farmer" method of information sharing. It brings scientists, farmers, and consultants together in a collaborative, 'co-learning' environment that enables farmers to learn and adapt integrated farming practices to local conditions." (SAREP 1999)

14. IPM built on the ecological principles in biological control (e.g., predator-prey relationships and population dynamics) in one major way: using cultural strategies to create habitat conducive to beneficial insects and degrade habitat amenable to pests. Organic growers had carried forward the tradition of fully integrating cover crops into their farming systems, relying on them as a major source of fertility and to moderate pest populations (Vos 2000). For an empirical study of organic practices and the difficulties they face approaching agroecological ideals, see Guthman 2000.

15. The official history of pesticide regulation is described in California Department of Pesticide Regulation 2001b. For helpful background information, see Stoll 1998 and California Environmental Protection Agency 2001.

16. On the creation and development of this massive database, see California Department of Pesticide Regulation 2000. It was created during a period when USEPA was developing risk-based approaches to regulation. It appears that the complex array of federal and state laws shaping pesticide restrictions created a situation in which the value of knowledge about pesticide use became greater than the political cost to DPR of remaining ignorant. For an overview of how they use PUR data to track compliance with the multiplicity of pesticide regulations in California, see California Department of Pesticide Regulation 2002. Pesticides are reported in terms of pounds of active ingredient (a.i.), but this does not include other "inert" materials (solvents, emulsifiers, etc.), some of which are hazardous. DPR explains the fluctuation in pesticide use by pointing to annual variations in pest problems, weather, acreage and types of crops planted, economics, and other factors. For a discussion of limitations associated with PUR data, see Kegley et al. 2000, p. 16. UC Davis hosts a workgroup dedicated to its study.

17. See Department of Pesticide Regulation 1995. DPR drew its statutory authority to encourage less environmentally hazardous pest-control practices from its previous institutional home at the California Department of Agriculture.

18. Center for Agricultural Partnerships 2002, p. iv.

19. I exclude PMA grants awarded to agricultural groups for only one year and those awarded to urban, school, and floriculture groups. The following agricultural PMA grants were for one year only and are therefore excluded from this analysis: strawberries, lettuce, poultry, alfalfa seed, cotton, roses, rice, containerized nursery products, and turkey. For an analysis of the PMA program, see Center for Agricultural Partnerships 2002.

20. Benbrook (1996) described how the accretion of US pesticide laws resulted in contradictions, and policy entrepreneurs exploited these tensions to pass the FQPA. Smart (1998) describes the political forces and policy negotiations that lead to its passage. On the implementation of FQPA, see Groth et al. 2001 and USEPA 2004.

21. On policy windows, policy entrepreneurs, and the coupling of issues, see Kingdon 1995, pp. 165–195.

22. Osmond 2002, p. 5.

23. Osmond 2002, p. 7. Although large in scope, the Neuse partnership was relatively modest in its goals. It extended Best Management Practices, which many California agroecological partnerships, such as BIFS, dismiss as insufficient to promote a shift toward systems thinking on the part of growers.

24. See case study 7 in NRC 1989. See also Glades Crop Care Inc. 2003.

25. Kellogg's use of this term appears to mean integrated with social values or socially desirable agriculture. It does not refer to the National Research Council's report.

26. Berkenkamp and Mavrolas (2003) describe the individual projects. Fisk et al. (1998) describe Kellogg's theory of agricultural change.

27. Dlott et al. (1996) were the first to write about the BIOS partnership model. Schaffer (1997) built on that work and wrote a guide for reproducing elements of BIOS. Pence (1998) wrote a depth ethnography of BIOS and the Lodi winegrape partnerships. Pence and Grieshop (2001) articulated traits of the model and tensions with traditional agricultural research and extension practices.

28. Pence 1998, p. 4.

29. For an analysis of forestry partnerships, see Wondolleck and Yaffee 2000.

30. For a tabulation of grants awarded to California's agroecological partnerships, see Warner 2004, pp. 146–147.

31. See Pence 1998.

32. See SAREP 2003. For analysis, see Warner 2004, pp. 158–160.

33. "Gray" literature documenting the effects of partnership on participating growers is assembled in Warner's (2004) methodological appendix.

34. Stone fruit growers have reduced organophosphate use, in part by switching to "softer products" as have almond growers, but organophosphate use has not declined as dramatically in this commodity. Stone fruit growers and their organizations have not invested the same degree of effort into partnership activities as have almond growers. They also have to cope with greater cosmetic concerns.

35. Epstein et al. 2000, 2001; Epstein and Bassein 2003; Zhang et al. 2004; Elliott et al. 2004.

36. Winegrapes and almonds have comparable acreages in California. Winegrapes used less than 10 percent the organophosphates of almonds (Elliot et al. 2004, table 1A).

37. Nick Frey of the Sonoma winegrape partnership reports that winegrape pesticide use has declined in his county every year since 1997, even as grape acreage has increased by 50 percent during this period. Materials under FQPA review decreased between 1999 and 2001, and the acreage treated with these materials declined to 49 percent (Sonoma County Grape Growers Association 2003b; also documented in Campos and Zhang 2003 and in Campos and Zhang 2004, 2005). The efforts of the Central Coast Vineyard Team are documented in McDavit and O'Connor 2004.

Chapter 4

1. This narrative is based on the following sources: interviews with Karl Kupers, Tom Platt, and Chris Feise in August 2000; USEPA 1997c and 1998; Lucido 1999; Donovan 1999; AgHorizons Team 2000; Roberts 2001. Karl Arne and Diana Roberts also contributed.

2. No-till farming does not cultivate the soil at all. Ridge-till farming only ploughs a few inches on the top of ridges to prepare a small seedbed. Ridge-till farming is not used in the Pacific Northwest for grain crops.

3. Sources of information on the Dakota Lakes Farm: Donovan 1999; Dwayne Beck, personal communication.

4. Source of this quotation and the next: Donovan 1999.

5. For a discussion, see National Research Council 1993, pp. 322–325.

6. Allan Savory (1988) developed the Holistic Resource Management approach to agriculture, wildlife management, and social equity in Zimbabwe (formerly Rhodesia). Dwayne Beck drew from the HRM approach's integration of human values into growers' decision making. The Chilean agronomist Carlos Crovetto Lamarca (1996) also integrated HRM into field crop systems. In the US, graziers and ranchers have used HRM more than other agricultural producers.

7. Described in USEPA 1997b.

8. See discussion in chapter 4 of Warner 2004.

9. Stephen Stoll (1998, p. 32) writes of late-nineteenth-century California: ". . . the men and women who took up irrigated lands to cultivate trees and vines did not choose to do so because farming was the only life they had ever known, nor because they identified virtue with work close to the soil. Fruit growers were more likely to see themselves as business people than as toilers; indeed, the people who settled crop districts from San Jose to San Diego often refused the title 'farmer.' Instead they referred to themselves as 'growers'—orchard capitalists— and they expected more from the fruit business than a bare living. As one promoter put it, they wanted 'farming that pays.'"

10. Richard Walker (2004) calls California's "the most perfect capitalist agriculture in the world." The Gold Rush of 1849 established economic ties and generated wealth that subsequently circulated from California city to countryside, and specialized monocultural production sprouted where growers thought local conditions favorable (Henderson 1999). Agrarian capitalism never competed here with other social relations or production methods, such as integrated family farming, sharecropping, or plantation.

11. Perennial crop growers and their organizations have deployed the agroecological partnership model with much greater enthusiasm than other growers, and the description that follows should be understood to refer primarily to them because the vast majority of my investigations were among them. This description of growers should be understood to apply foremost to perennial crop growers.

12. The first diffusion-of-innovation study was that of Ryan and Gross (1943). For a description of this tradition, see Rogers 1983. See also Ruttan 1996.

13. This study is reported in Brodt et al. 2004. The methodology is described in greater detail in Brodt et al. 2001. For a summary, see Warner 2004, pp. 194–199. Brodt and her colleagues posed 48 statements that allowed growers to rate their relative preference for an economic and personal satisfaction goal. They used a quantitative technique to analyze grower subjectivity known as Q methodology, an "inverted" form of factor analysis that "shifts the focus toward intercorrelations of people based on each individual's overall pattern of all traits tested for, and away from intercorrelations of individual traits." They asked each subject to rank their agreement with a whole series of questions that reveal their own internal frame of reference. They interviewed 40 growers in northern San Joaquin Valley, half participants in a partnership and half not. Each grower was asked to sort a set of 48 statements designed to express a belief, a general value, or a management goal. These statements were carefully worded so as to "force" the grower to rank his or her relative preference for an economic and a satisfaction factor. The researchers entered the results into the PQ method software that performed principal components analysis and a varimax rotation to identify a small number of heavily loaded factors, which they used to define these three management styles.

14. The threefold typology of Brodt et al. should not obscure the fundamental fact that all but a handful of growers are attracted to partnership activities primarily because they believe they can learn how to cut costs, as other surveys of walnut and prune growers have shown. These surveys were conducted by SAREP in association with the California Dried Plum and Walnut Boards, and Farm Advisor Joe Grant (Walnuts). The walnut survey results are from Ransom et al. 2003. The prune survey results are not yet published. The data are summarized and discussed in Warner 2004; see especially table 3.4 on page 200. At the time of my field research (2002–03), winegrape growers, and to a certain extent almond growers, had enjoyed profitable years that offered the financial opportunity for some to experiment with some new practices, but no growers would voluntarily adopt these methods if they lost money in the process.

15. See Carter et al. 1996.

16. Busch and Lacey 1983. See also Busch and Lacey 1986. For a summary of this early work on the sociology of agricultural scientists, see Buttel et al. 1990. The organization and management of LGU scientists and science activities is a specialized subfield but has produced a considerable amount of literature. Early work by Isao Fujimoto and colleagues (Fujimoto and Fiske 1975; Fujimoto and Kopper 1975) led to considerable controversy, although their findings hardly seem controversial today.

17. Recall figures 2.1 and 2.2, which interpret these shifts.

18. The USDA's Agricultural Research Service (ARS) also hosts scientists conducting basic research, and they may have even more prestige than UC agricultural scientists because they conduct fundamental, long-term, high-risk research that may not produce practical knowledge. This mission is not compatible with the applied nature of partnership research, and thus only three ARS scientists have contributed to California's partnerships. Only one ARS scientist led a partnership, and she reported it to be highly unusual for her position.

19. The University of California has historically had a greater role for specialists than other state extension systems (Learn, Lyons, and Meyers 1987). Formerly they were based in county offices and provided specialized assistance to Farm Advisors, but they are now based on campus, required to have a PhD, and evaluated more by their scientific publications than their extension efforts.

20. According to California law, anyone who provides pest-control recommendations concerning any agricultural use, offers himself as an "authority" on any agricultural use, or solicits services or sales for agricultural use outside a fixed place of business must hold a PCA license. See California Department of Pesticide Regulation 2001a. Growers can, however, still apply pesticides on their own fields and orchards without a PCA recommendation.

21. My demographic information about PCAs comes from the California Agricultural Production Consultants Association's member surveys, which also indicate that about half of PCAs work in agriculture (CAPCA 1999). There were 4,300 PCA license holders in 1977, and this number has remained stable for 25 years. According to Mac Takeda of DPR (personal communication), there were 4,418 licensed PCAs in the state as of June 26, 2003.

22. Agrochemical manufacturing and sales companies had sprung up across California in the early twentieth century to provide pesticides for specialty-crop agriculture (Stoll 1998). Specialized crops required specialized services, so salesmen would travel the countryside offering them specialized agrochemicals. Some of these salesmen were highly unscrupulous, and few had any scientific training. In 1970, spurred by *Silent Spring*, State Senator Anthony Beilenson proposed legislation that would prohibit salesmen from making pesticide recommendations, which would sever the link between prescription and sales. This bill was sidetracked by the agrochemical industry, but eventually a weakened S.B. 1020 passed in 1972, requiring pesticide salesmen to take a nominal test and receive a PCA license. For a summary of the bill and its subsequent impact from an industry perspective, see CAPCA 2003. For a description of how the "pesticide mafia"

gutted the original intent of the bill, see Van den Bosch 1978. Testing requirements for PCA licenses have continued to expand, and every new PCA must have a bachelor's degree in biological or agricultural science and must pass several written exams. These entrance requirements have, over time, reduced the number of quacks, or "old boys," who promoted spraying based only "on the calendar," thus frequently over-prescribing.

23. The PCA industry has not been able to standardize the market for its services and establish consistent expectations for behavior by its members. Many affiliated PCAs point to the improved professional behavior of the industry over the past two decades. For an analysis of the professionalization process, see Larson 1977. For a review of the ethical responsibilities of scientific professionals, see Ravetz 1988.

24. For an analysis, see chapter 4 of Warner 2004.

25. Forty-nine of the 257 UC Farm Advisors participated in 24 partnerships, a participation rate of 19.0 percent. Each Farm Advisor participated in only one or occasionally two partnerships, the one exception being San Joaquin pomology Farm Advisor Joe Grant, who participated in five. Five Farm Advisors were Principal Investigators on BIFS and other grants, which dominated the activities of the Farm Advisors for the duration of the grant. Farm Advisors played a prominent role in leading twelve other partnerships. Half of these were PMA-funded partnerships in which commodity organizations were awarded the grant but depended on the Farm Advisor to organize the research and outreach activities.

26. The regional IPM Advisors occupy an intermediary niche between Extension Specialists and Farm Advisors, and the UC IPM program has repeatedly had to negotiate their relationship to UCCE (Lyons 2004). UC leadership periodically attempts to reclassify IPM Advisors as generic Farm Advisors, which UC IPM directors have resisted actively and successfully.

27. Twenty-four of the 149 Extension Specialists (16 percent) participated in partnerships, and five of them were principal investigators. The percentage of Extension Specialists who participated in partnerships is slightly less than the percentage of Farm Advisors, but an equal number were principal investigators.

28. Source: presentation by Richard Waycott at conference of Almond Board of California, December 6, 2002.

29. Thirty partnerships included UC personnel explicitly as partners. The two exceptions were Jeff Dlott's doctoral studies of farms owned by the California Clean Growers farmers and his consulting work on the Code of Sustainable Winegrowing Practices. In each of these cases, Dlott used UC-generated knowledge, even though UC scientists were not formally enrolled.

30. Fourteen scientists participated in fourteen partnerships, but two participated in six each, driving the average number of partnership per scientist up to two. Stephen Welter and Nick Mills, both scientists at UC Berkeley, have worked respectively on pheromone mating disruption in tree crops and the biological control opportunities that emerge after broad spectrum insecticides are removed.

31. For more on the historical development of commodity organizations, see Friedland and Haight 1985, Pincetl 1999, and Saker 1990. There are now 53 commodity organizations authorized by the state of California and 10 authorized by the federal government. In California, 41 serve crop-plant growers and 12 serve seafood, forestry, floriculture, and root-stock agriculture (California Department of Food and Agriculture 2000). All ten federal marketing orders in California serve perennial crops. Growers founded and support these to help with their specialized production and marketing needs. All are semi-public corporate groups financed by a mandatory tax imposed by a majority vote of the growers of a specific geographic area. All but a few have the primary purpose of promoting the marketing of a particular agricultural product through promotion, advertising, and the imposition of quality standards. Most are also authorized to fund production research, although usually only a tiny fraction of organizational budgets are devoted to this. A few California commodity organizations are authorized for research exclusively ("councils" and "commissions").

32. Bunin 2001.

33. The Almond Board of California, for example, dispenses half a million dollars per year in production research grants. The Walnut Board dispenses a roughly equal amount, even though the value of the state's walnut crop is only one-third the value of its almond crop.

34. The almond industry is large, growing, and high-profile, and its practices affect multiple environmental media. The almond harvest generates dust throughout the Central Valley. In the winter, the burning of almond prunings harms air quality. Any pesticide used by the almond industry is going to attract a lot of attention, as diazinon has. As towns and cities along Highway 99 sprawl onto farmland, almond growers are increasingly confronting urban interface issues. The almond industry is so large that it had to take on these new responsibilities.

35. See Marthedal 2001.

36. Examples include the Central Coast Vineyard Team, the Napa Sustainable Winegrape Growers, the Sustainable Cotton Project, and the Ukiah Valley IPM Pear Growers.

37. See USEPA 1997a.

38. See Andrews 1999. Some states publicly complained about the USEPA's aggressive stance impinging on states' prerogatives, but at the same time, they have privately welcomed its initiatives by raising compliance standards and strengthening their negotiating positions.

39. Examples of agencies providing small grants in California: Resource Conservation Districts and USDA programs, National Resource Conservation Districts, the Cooperative State Research and Extension Service, and Western Sustainable Agriculture Research and Education. A few agencies awarded large grants to one or two partnerships (CalFed, the State Water Quality Control Board, or the California Department of Food and Agriculture).

40. For an excellent analysis of CPAI, see Feise and Lovrich 2003.

Chapter 5

1. Source of data: UC-IPM 2002. The so-called NOW was first found feeding on damaged navel oranges in Arizona.

2. For specific data on regional changes in almond acreages, see Warner 2004, pp. 256–258.

3. Scientists had observed that the new mechanical shakers left more nuts on the trees, although uneven moisture levels in the tree also contributed. Scientists recognized harvest date could also play a role in NOW control. Nonpareil, the most commonly planted variety, ripens early and has a soft shell that does not always seal properly. It is however, preferred by handlers because of its mild taste and ease of shelling. Many seasons Nonpareil are ready to harvest about the same time the second generation of NOW are laying eggs for a third generation, which can quickly hatch during the hot summer days and attack the vulnerable nuts. By harvesting Nonpareil as soon as its nuts ripen, growers can spirit them away from the NOW. Later pollinating varieties are generally less susceptible to pest damage. Conducting two harvests—one for early and one for later varieties—costs more, but is cheaper than an extra insecticide treatment at hullsplit. It requires additional monitoring, a harvest strategy flexible enough to take advantage of biological time, and a desire to reduce pesticides.

4. This was part of a larger IPM funding initiative the USDA launched to rehabilitate public perception of the agency after Congress transferred pesticide regulatory responsibility to the USEPA. See Almond Board of California 1998.

5. See Klonsky, Zalom, and Barnett 1990.

6. *Goniozus legneri* attacks only the NOW and the carob moth (*Ectomyelois ceratoniae*). That led Legner to conclude that the NOW had originated in southern South America.

7. See Legner and Gordh 1992, pp. 2158–2159.

8. Bentley et al. 2001.

9. For a discussion of adaptive management, see Röling and Wagemakers 1998.

10. For discussions of commodity systems analysis, see Friedland, Barton, and Thomas 1981, Friedland 1984, and Friedland 2001.

11. For a more detailed programmatic analysis of these five chief strategies, see chapter 4 of Warner 2004. A caveat is needed about the numbers organized into categories of practices and chief strategies. Sources of information for these tables are interviews of partnership leaders, and partnership documentation and reports. The numbers represent at a minimum that a practice was tried by at least one grower in the partnership. In most cases, most growers tried most practices, and many adopted them.

12. A few California efforts that used the term "partnership" were, in fact, essentially technology-transfer activities that found the language of "partnership" useful for fund-raising purposes. They were included in this study primarily for purposes of comparison.

13. Schaffer's (1997) guidebook describes how CAFF set up BIOS partnerships in various counties and strategies for enrolling growers. Roughly half were recruited by independent PCAs and Farm Advisors, and half through meeting announcements (Pence 1998). According to Schaffer's book, BIOS expected growers to (1) wish to reduce the use of chemicals, especially hazardous pesticides, (2) dedicate 15–30 acres of almonds for BIOS experimentation paired with a "conventional" orchard block of equal size, and (3) gather data and share information learned in this process with management team members, other participating growers, and the broader agricultural community at field days.

14. Partnership leaders perceive pest-control advisors as shaping existing farming practices, although they hold a range of opinions on their relative importance as an audience for partnership activities. This diversity of views is reflected in the variety of strategies for enrolling PCAs. In some cases the partnership's management team enrolls an independent PCA as an applied scientific expert, and in others, the PCA is enrolled along with the grower. In twelve partnerships some blend of these occurred. The management team PCA can model agroecological pest-management skills, but may be perceived as a threat by affiliated PCAs. Partnership leaders have not, in general, fully enrolled the PCAs of partnership growers. Generally, leaders report that these PCAs have kept their distance.

15. For growers in the four local winegrape partnerships, joining a partnership simply added a dimension to existing relationships of cooperation. See Warner in review (b).

16. For a discussion of the environmental complex in agroecology, see Gliessman 1998, pp. 167–175.

17. "Pesticide" refers to any material used to kill pests, be they insects, diseases, weeds, or vertebrates. Partnerships have focused chiefly on insects, and secondarily on weed pests. Plant disease management uses less hazardous materials, and fewer opportunities to deploy agroecological techniques exist to treat them.

18. Field-crop ("Westside BIFS"), rice, and dairy BIFS partnerships focused more attention on soil and nutrient management.

19. For an analysis of the changes resulting from organophosphate removal in 15 California crops, see the appendix to Metcalfe et al. 2002. See also Steggall 2003.

20. Some growers and consultants use the term "cover crop" to refer to resident orchard vegetation. This may on the whole offer more advantages than disadvantages to the grower, especially in terms of water infiltration and dust (and mite) control, but for the purposes of this study I do not consider this to be a cover crop. I do not consider it an agroecological strategy because it is not a management action.

21. Daane and Costello (1998), who investigated the relationship between cover crops and leafhopper damage levels, concluded that cover crops attenuated late season vine vigor, which indirectly reduced pest pressure.

22. Even though the University of California has promoted the use of cover crops for decades, it has not been a research priority. According to Fred Thomas:

"If you want to do it right you have to do ten to twenty varieties, and we are talking four replications, and sampling, and we are talking a long term, $600,000 four year grant to do the right research to have good data." This is an example of the kind of research that could be a tremendous help for growers interested in agroecological strategies, but is perceived to be too repetitious by UC scientists. It is too applied to be of scientific interest.

23. Ingels et al. 1998 is the best source of information on cover crops in California.

24. Ingels and Klonsky (1998) informally surveyed UCCE Farm Advisors, asking them to estimate the percentage of all vineyard acreage cover cropped. North Coast winegrape growers were more likely to use cover crops.

25. Some vineyards on the Central Coast are planted on hillsides particularly vulnerable to erosion, and the Central Coast Vineyard Team has emphasized the importance of cover crops to protect soils.

26. This refers to the "Westside BIFS" in western Fresno County. See Mitchell et al. 1999 and Mitchell 2001.

27. CAFF 1995.

28. ". . . a farmer's role is not solely to tend the crop and defend it from pests, but also includes being a resource manager: building the soil, enhancing habitat, nurturing beneficial organisms, and monitoring orchard interactions. In this way they derive economic benefit from the products of naturally occurring processes, such as nutrient cycling and biological control." (CAFF 1995, p. 11)

29. The prune partnership and Sun-Maid BMP manuals consist of decision rules for individual practices, and do not relate the benefits of farming systems integration.

30. Jenny Broome, then a scientist working for the California Department of Pesticide Regulation, recommended that the group develop a point system for monitoring the adoption of practices.

31. The process of creating the Positive Points System forced them to be specific about what they really meant by sustainable farming, and to evaluate the relative importance of different agroecological practices. They had to arrive at an objective number for evaluating the relative sustainability of their activities—pest, soil, and water management, and viticultural, wine quality, and continuing education practices—as applied to the diverse geographic contexts of the Central Coast. As a management tool, the PPS provides an objective guide for growers' annual inventory of their practices. Growers report their scores to the Central Coast Vineyard Team, and they usually increase during the first few years use because it asks growers to consider dimensions of their farming system to which they may have previously given little attention. The CCVT has continued to update the PPS. See Central Coast Vineyard Team 2003 and McDavit and O'Connor 2004.

32. The workbook is presented in "workshops" at which growers/managers score their performance in six thematic areas on 109 questions on a point scale of 1–4, with 4 representing greater sustainability. Ohmart credits the Central

Coast Vineyard Team for the idea of developing a tool and the Farm*A*Syst program in Wisconsin for the concept of a four-category evaluation suggesting a "growers' action plan." For more on the accomplishments of the Lodi winegrape partnership, see Ohmart 2001 and chapter 7 below. The workbook is designed to challenge and stimulate Lodi growers regardless of their present practices, but also to provide recognition for the district and serve as a basis for future third-party certification.

33. The diversity of crop and animal yields from agroecological farming in developing countries increases food security in a subsistence context. See Holt-Giménez 2002. On cropping diversity and risk management, see Uphoff 2002.

34. Bentley et al. 2001 is one of the few examples of incorporating risk into the evaluation of BIOS or BIFS practices.

Chapter 6

1. See Division of Agricultural Sciences 1978.

2. The states included Washington, Oregon, California, and Colorado. Sources of data on CAMP: Calkins and Faust 2003, Welter et al. 2005.

3. Most previous work has concentrated only on the individuals or exclusively on institutions and social structures. Social scientists generally deploy either micro- or macro-analytic techniques to interpret society. Both of these methodological approaches fail to convey the relational dimension of partnerships. Macro approaches, such as political economy, reveal the structures and institutions in society and how they shape the way people think about the world, but too often present reality as already constituted and immutable. They also tend to filter out the agency of individuals seeking to effect change within their world. Micro approaches tend toward positivism and behavioralism, and fail to adequately reveal the importance of social relationships and associations.

4. My work does not deploy a full actor-network-theory methodology. For an introduction to actor-network theory, see Latour 1987 and Callon and Law 1989. As developed, ANT is really more a method to flesh out how actors build networks of knowledge, practice, and political influence than a theory (Latour 1999). ANT offers a sprawling assemblage of methods, but there are at least two other currents in network analysis: sociometric approaches (Wasserman and Faust 1994) and the actor-oriented approach (Long and Long 1992). These approaches to networks do not give accord sufficient voice to the non-human, nor adequately consider the epistemological controversies that emerge from hybrid nature-society networks. Latour in particular has studied networks of scientific knowledge and the strategies of various social actors to use knowledge to influence the behavior of others. His work implies that all scientific enterprises have had a politics, that is, have managed people and institutions—as well as biological and technological objects—through persuasion. For another example of ANT methodology in the agrofood system, see Whatmore 1997.

5. See Hassanein 1999.

6. I base my analysis on interviews with participants, and on qualitative data from and original sociograms drawn by partnership leaders. All 32 partnership leaders created original sociograms for me, which I used as primary data for understanding how actors were configured and their relative importance to knowledge exchange. For all eight sociograms presented here, I standardized the actors into six categories (growers, pest-control advisors, Farm Advisors, scientists, growers' organization, commodity organization), but added other actors as appropriate for each partnership. For more information on the derivation of these sociograms, see the methodological appendix to Warner 2004.

7. The ABC later became the first and only commodity organization with an Environmental Committee to address the environmental impacts of its production practices. CAFF also tried to influence the research agenda of the Production Research Committee, which had historically funded pesticide-oriented and IPM-oriented pest-management research, but there is little evidence that it was successful.

8. It was authored by two UC IPM Advisors and three almond Farm Advisors (Pickel et al. 2004).

9. Historically, the ABC had closer ties to the Almond Hullers and Processors than to growers because they collect the state authorized "tax" on the almonds and turn it over to fund the ABC. Before 1999, the ABC did not even have a list of the growers in the state because their names were guarded closely by the approximately 100 economic entities that processed almonds. The ABC also worked closely with the Almond Hullers and Processors Association to lobby for the almond industry, an activity prohibited by law for commodity boards.

10. See Kimbrell 2002. Most troubling to the PMA partners was this passage: ". . . the industrial approach to almond production comes at a staggering cost to the environment. According to the EPA, almond production is among the top polluting agricultural industries in the region. Through the use of massive chemical inputs, mechanical harvesters, and enormous quantities of imported water, California almond corporations weak havoc on the land. Among the greatest threat of industrial almond production is the widespread use of toxic pesticides, particularly the nerve toxin Diazinon. In recent years, for example, the levels of almond pesticides have skyrocketed above allowable limits in all of the major tributaries to the San Francisco Bay." (Kimbrell 2002 p. 172) PMA partners were understandably upset by this. Kimbrell and his contributors created a false binary contrast between the agrarian, sustainable, organic vision and the "fatal" vision of industrial agriculture that ignored the years of work BIOS and the almond PMA had done.

11. For a discussion of sustainability in the winegrape industry, see Warner in review (b). The process began informally during the middle years of the twentieth century as Napa and California wine producers recognized the opportunity to enhance their reputation of quality by providing more information to consumers. The geographic branding of winegrape production has been the most important cooperative strategy of grape growers and winemakers for nearly 40 years in California. Lapsley (1996) and Conaway (1990) provide the best overview of

how this cooperative approach was pioneered by Napa winegrape growers and vintners. It has been copied to varying degrees by other premium production regions.

12. Of all the industry leaders, Robert Mondavi Winery has played a singular role in constructing association between wine and environmental quality. It has done this on its own "estate" lands with contract growers. Upon his return from Europe in the 1960s, Robert Mondavi launched an incentive system for quality grape growing, and later developed an education program for growers that would have them taste the range of wine quality as it varied according to conditions of production (Mondavi 1998). In the 1990s, this effort became formalized as "quality improvement groups" in regions outside Napa, and this laid early groundwork for partnerships in the Lodi and Central Coast regions. Robert Mondavi Winery management facilitated growers addressing winegrape quality, which came to include environmental quality as well. Winegrape growers and vineyard managers in Napa and elsewhere confirm the importance of this winery's leadership. Mondavi shifted from an IPM and piecemeal approach to environmentally friendly agriculture to practicing sustainable agriculture on winery-owned lands in 1990. The recent sale of Mondavi clouds the future of these commitments.

13. See Conaway 1990, Conaway 2002, Friedland 2002, and Warner 2006b.

14. One participant in NSWG said ". . . a system where the workers understand sustainability is more sustainable than when it's just the bosses in the coffee shop and it's an intellectual thing." Another NSWG participant said "the person handling the vine has to be the one who understands the principle behind it, and is able better to implement it then as a result."

15. See Friedland 2002.

16. In 2003, Robert Mondavi Winery, Beringer-Blass, Diageo, E. & J. Gallo, and Fetzer collectively farmed or contracted vineyard management for 18,000 acres, and buy from 19,000 acres, consuming 42 percent of the region's 86,5000 acres.

17. Sulfur accounts for roughly one-third of the PUR-reported pesticide use in California, and 66 percent of its use is on grapes. From 1997 to 1999, of all grape sulfur drift incidents that took place in California, 35 percent were in the North Coastal Counties (with 26 percent of the winegrape acreage) and 11 percent were in the Central Coast Counties (with 18 percent of the acreage), compared with 43 percent in the south San Joaquin Valley (with 30 percent of the acreage). The number of drift complaints has increased over time, due to increased acreages but also these regions' extensive winegrape/residential interface. Of all California perennial crops, winegrapes have the second highest acreage, and are thus the leading perennial crop user of persistent, or pre-emergent, herbicides under FQPA review. The DPR has discussed further restrictions on sulfur, and supported the winegrape industry working cooperatively to reduce complaints about sulfur and the use of these materials in general. See Browde 2002.

18. This synergy is described in greater detail in Warner in review (b).

Chapter 7

1. Lapsley 1996.

2. The UC Davis Department of Viticulture and Enology provided early, critical work to help growers recognize that quality winegrapes required different management practices than table or raisin grapes. It also conducted essential research on varietals and their role in quality wine production.

3. Elizabeth Barham (2003) contrasts US and European understandings of "terroir."

4. For more on the dynamics of the California winegrape industry, see Warner in review (b).

5. The grape leaf hopper (*Erythoneura elegantula*) is the primary insect pest for Lodi grape growers, and the myramid wasp *Anagrus epos* can provide biological control, although it does not do so consistently. Initially growers and researchers were enthusiastic about the possibility of using French prune trees as hosts for alternate, counter-seasonal prey for the *Anagrus*, but they discovered that to be effective, growers would have to devote an equal area to prune trees as winegrapes, an impractical proposal for modern farming. On biocontrol of leaf hoppers, see Murphy et al. 1996. Researchers had also hoped that cover crops could be an alternate host for other beneficial insects, but they determined instead that cover crops had an indirect effect on pest pressure through competing with vines for nutrients. Even though cover crops increase water use, growers have determined their use resulted in a significant reduction in mites. On cover crop/vine/insect pest relationships, see Costello and Daane 1998 and Daane and Costello 1998. Growers express the most enthusiasm for deficit irrigation research demonstrating that by moderately stressing vines by withholding irrigation water for a few weeks during the early summer months improves color and flavors later in the season. Lodi growers now typically aim to harvest 5–7 tons per acre, a marked reduction from previous decades.

6. This periodization and most of the information in this section is taken from a conference *Twelve years of sustainable viticulture in Lodi*, held March 24, 2004, summarized in Ohmart 2004.

7. Ohmart 1998.

8. He said this at the "Twelve Years of Sustainable Viticulture in Lodi" meeting, held March 24, 2004.

9. Arounsack et al. (2004) have studied decreases in FQPA priority herbicides among Lodi BIFS growers. Organophosphate use in among Lodi winegrape growers has dropped to virtually nil. This appears to be chiefly because: alternative pesticides have been developed (manufacturers do create new products for larger California commodities like winegrapes); growers have realized organophosphates are not necessary to grow quality winegrapes; for years the commission has discouraged their use. The Vine Mealybug is beginning to infest some vineyards in the district, and this has resulted in increased Lorsban use (an organophosphate) because no softer products yet exist. Ohmart's analysis of

insecticide and herbicide use among BIFS growers confirms the findings of others that weather has had the greatest impact (Zhang et al. 2004).

10. Ohmart's (1998) data provides the baseline, and Dlott (2004) describes the changes as a result of 5 subsequent years of commission efforts. They used virtually identical survey questions. All data is reported by growers and subject to their bias. The data in table 7.1 is in response to the question "As a result of the commission's efforts I. . . ." Monitoring more frequently went from 66 percent (1998) to 78 percent (2003). Monitoring more systematically jumped from 49 percent to 70 percent. Increased monitoring of beneficial insects stayed roughly the same. The growers were asked "Do you spend more time monitoring per trip to vineyard?" The number reporting spending more time monitoring increased from 55 percent to 63 percent. In 1998, Ohmart received 288 of 608 surveys mailed (a 47 percent response rate). In 1993, Dlott received 307 of 712 (a 43 percent response rate).

11. See Bisson et al. 2002.

12. On other eco-label initiatives, see SAREP 1998, Central Coast Vineyard Team 2000, and Warner in review (b).

13. On the Code's initial progress, see Dlott 2004.

14. To help rationalize pesticide regulatory decision making, the USDA created a program to help growers and their organization formalize crop-specific knowledge about pests. All of California's perennial-crop commodity organizations with partnerships have developed "Pest Management Strategic Plans."

Chapter 8

1. On endocrine disruptors, see Colborn, Dumanoski, and Meyers 1996. This case study is based on interviews with Deana Sexson, Sarah Lynch, Carolyn Brickey, and Jeff Dlott, and supplemented with data from Gordon 2002 and Lynch et al. 2000.

2. On the IPM continuum and bioIPM, see Benbrook 1996.

3. These five points are from Lynch et al. 2000.

4. See Benbrook 1996. On the toxicity index, see Benbrook et al. 2002 and Lynch et al. 2000.

5. On Colorado potato beetle area-wide management, see Sexson and Wyman 2005.

6. Information taken from Morse 2004.

7. For more on this, see Land Institute 2003 and Glover 2005.

8. On the Fund for Rural America, see Marshall 2000.

References

AgHorizons Team. 2000. The Wilke Project: Field Day and Progress Report. Washington State University Cooperative Extension.

Allen, Patricia. 2004. *Together at the Table: Sustainability and Sustenance in the American Agrifood System*. Pennsylvania State University Press.

Allen, Patricia, ed. 1993. *Food for the Future: Conditions and Contradictions of Sustainability*. Wiley.

Allen, Patricia, Debra Van Dusen, Jackelyn Lundy, and Stephen Gliessman. 1991. Integrating Social, Environmental and Economic Issues in Sustainable Agriculture. *American Journal of Alternative Agriculture* 6, no. 1: 34–39.

Almond Board of California. 1998. Years of Discovery: A Compendium of Research Projects 1972–1998. Almond Board of California.

Altieri, Miguel. 1989. Agroecology: A New Research and Development Paradigm for World Agriculture. *Agriculture, Ecosystems, and Environment* 27: 37–46.

Altieri, Miguel. 1998. Ecological Impacts of Industrial Agriculture and the Possibilities for Truly Sustainable Farming. *Monthly Review* 50, no. 3: 60–71.

Altieri, Miguel. 2002. Agroecology: The Science of Natural Resource Management for Poor Farmers in Marginal Environments. *Agriculture, Ecosystems, and Environment* 93: 1–24.

Andrews, Richard. 1999. *Managing the Environment, Managing Ourselves*. Yale University Press.

Arounsack, S. Steve, Minghua Zhang, and Janet Broome. 2004. Understanding Winegrape Weed Management Practices of a Biologically Integrated Farming System in San Joaquin County, California. Poster, Plant and Soil Conference, American Society of Agronomy, California Chapter.

Aspelin, Arnold. 2000. Pesticide Usage in the United States: Trends During the Twentieth Century. http: //www.pestmanagement.info/pesticide_history, accessed January 19, 2004.

Aspelin, Arnold, and Arthur Grube. 1999. Pesticides Industry Sales and Usage: 1996 and 1997 Market Estimates. USEPA Document 733-R-99-001.

Barham, Elizabeth. 2003. Translating Terroir: The Global Challenge of French AOC Labeling. *Journal of Rural Studies* 19: 127–138.

Bell, Michael Mayerfeld. 2004. *Farming for Us All: Practical Agriculture and the Cultivation of Sustainability.* Pennsylvania State University Press.

Benbrook, C. 1996. *Pest Management at the Crossroads.* Consumers Union.

Benbrook, C., D. Sexson, J. Wyman, W. Stevenson, S. Lynch, J. Wallendal, S. Diercks, R. VanHaren, and C. Granadino. 2002. Monitoring Progress on Reducing Reliance on High-Risk Pesticides in WI Potato Production. *Journal of Potato Research* 79: 183–199.

Bentley, Walt, Lonnie Hendricks, Roger Duncan, Cressida Silvers, Lee Martin, Marcia Gibbs, and Max Stevenson. 2001. BIOS and Conventional Almond Orchard Management Compared. *California Agriculture 55,* no. 5: 12–19.

Berkenkamp, JoAnne, and Pam Mavrolas. 2003. Changing Attitudes, Changing America's Food System. W. K. Kellogg Foundation.

Bisson, Lind, Andrew Waterhouse, Susan Ebeler, M. Andrew Walker, and James Lapsley. 2002. The Present and Future of the International Wine Industry. *Nature* 418, 8 August: 696–699.

Bohm, David. 1993. Last Words of a Quantum Heretic. *New Scientist* 137, 27 February: 42.

Bosso, Christopher. 1987. *Pesticides and Politics: The Life Cycle of a Public Issue.* University of Pittsburgh.

Brodt, Sonja, Karen Klonsky, and Laura Tourte. 2001. Farmer Goals and Management Styles: Implications for Adoption of Sustainable Agriculture. Presented at meeting of Western Economics Association, San Francisco.

Brodt, Sonja, Karen Klonsky, Laura Tourte, Roger Duncan, Lonnie Hendricks, Clifford Ohmart, and Paul Verdegaal. 2004. Influence of Farm Management Style on Adoption of Biologically Integrated Farming Practices in California. *Renewable Agriculture and Food Systems* 19, no. 4: 237–247.

Broome, Janet, W. Settle, Robert Bugg, Marcia Gibbs, and Clifford Ohmart. 1997. Biologically Integrated Farming Systems: Approaches to Voluntary Reduction of Agricultural Chemical Use. Presented at meeting of Society of Environmental Toxicologists and Chemists, San Francisco.

Browde, Joseph. 2002. Pest Management Alliance Project Final Report. California Association of Winegrape Growers.

Browne, William. 1988. The Fragmented and Meandering Politics of Agriculture. In *U.S. Agriculture in a Global Setting,* ed. M. A. Tutwiler. Resources for the Future.

Browne, William. 1995. *Cultivating Congress: Constituents, Issues, and Interests in Agricultural Policymaking.* University of Kansas Press.

Browne, William. 1998. *Groups, Interests, and US Public Policy.* Georgetown University Press.

Browne, William, Jerry Skees, Louis Swanson, Paul Thompson, and Laurian

Unnevehr. 1992. *Sacred Cows and Hot Potatoes: Agrarian Myths in Agricultural Policy, Growth & Change.* Westview.

Bunin, Lisa. 2001. Organic Cotton: The Fabric of Change. Dissertation, University of California, Santa Cruz.

Busch, Lawrence, and William Lacy. 1983. *Science, Agriculture, and the Politics of Research.* Westview.

Busch, Lawrence, and William Lacy. 1986. *The Agricultural Scientific Enterprise: A System in Transition.* Westview.

Buttel, Fredrick. 1993. Ideology and Agricultural Technology in the Late Twentieth Century: Biotechnology as Symbol and Substance. *Agriculture and Human Values* 10, no. 1: 5–15.

Buttel, Fredrick. 2001. Land-Grant/Industry Relationships and the Institutional Relationships of Technological Innovation in Agriculture. In *Knowledge Generation and Technical Change,* ed. S. Wolf and D. Zilberman. Kluwer.

Buttel, Fredrick. 2005. Ever Since Hightower: The New Politics of Agricultural Research Activism in the Molecular Age. *Agriculture and Human Values* 22, no. 3: 275–283.

Buttel, Fredrick., Olaf Larson, and Gilbert Gillespie. 1990. *The Sociology of Agriculture.* Greenwood.

CAFF (Community Alliance with Family Farmers). 1995. *BIOS for Almonds.*

CAFF. 2004. CAFF History.

California Association of Winegrape Growers and The Wine Institute. 2003. Code of Sustainable Winegrowing Practices Self-Assessment Workbook.

California Department of Food and Agriculture. 2000. Resource Directory 2000.

California Department of Pesticide Regulation. 1995. Pest Management Strategy for the Department of Pesticide Regulation.

California Department of Pesticide Regulation. 2000. DPR Pesticide Use Reporting: An Overview of California's Unique Full Reporting System.

California Department of Pesticide Regulation. 2001a. Laws and Regulations Study Guide.

California Department of Pesticide Regulation. 2001b. The California Story, a Guide to Pesticide Regulation in California.

California Department of Pesticide Regulation. 2002. Summary of Pesticide Use Report Data 2001.

California Environmental Protection Agency. 2001. The History of the California Environmental Protection Agency.

Calkins, Carrol, and Robert Faust. 2003. Overview of Areawide Programs and the Program for Suppression of Codling Moth in the Western USA Directed by the USDA-ARS. *Pest Management Science* 59, no. 6–7: 601–604.

Callon, Michel, and John Law. 1989. On the Construction of Socio-Scientific Networks: Content and Context Revisited. In *Knowledge and Society,* ed. L. Hargens, R. Jones, and A. Pickering. JAI.

Campbell, David. 2001. Conviction Seeking Efficacy: Sustainable Agriculture and the Politics of Co-optation. *Agriculture and Human Values* 18: 353–363.

Campos, Jennifer, and Minghua Zhang. 2003. On Farm Innovation: Identifying Farmer Innovations of Low Risk Pest Management Using PUR. Presented to PUR Workgroup, October 24.

Campos, Jennifer, and Minghua Zhang. 2004. Progress Towards Reduced-Risk Pest Management. *Practical Winery & Vineyard*, March/April: 1–6.

Campos, Jennifer, and Minghua Zhang. 2005. Pesticide Use in California Winegrapes Since the Enactment of the Food Quality Protection Act. Poster for the Agricultural Geographic Information System Laboratory, University of California, Davis.

CAPCA (California Agricultural Production Consultants Association). 1999. Pest Control Advisor Demographic Profile.

CAPCA. 2003. CAPCA History.

Carpenter, Stephen, Nina Caraco, David Correll, Robert Howarth, Andrew Sharpley, and Val Smith. 1998. Nonpoint Pollution of Surface Waters with Phosphorus and Nitrogen. *Issues in Ecology* 3: 16.

Carroll, John E. 2005. *The Wisdom of Small Farms and Local Food: Aldo Leopold's Land Ethic and Sustainable Agriculture*. New Hampshire Agricultural Experiment Station.

Carson, Rachel. 1962. *Silent Spring*. Houghton Mifflin.

Carter, Harold, Ray Coppock, Marcia Krieth, Ivan Rodriguez, and Stephanie Weber Smith. 1996. Voices of California Farmers: Effects of Regulations. University of California Agricultural Issues Center.

Center for Agricultural Partnerships. 2002. Evaluation and Recommendations for California's Department of Pesticide Regulation's Pest Management Alliance Program. California Department of Pesticide Regulation.

Central Coast Vineyard Team. 2000. Exploring Environmental Labeling and Certification for Sustainable Agriculture Programs.

Central Coast Vineyard Team. 2003. Positive Points System.

Central Valley Regional Water Quality Control Board. 2002. 2002 CWA Section 303(d) List Of Water Quality Limited Segment.

CIAS (Center for Integrated Agricultural Systems). 1998. Perspectives: A Background Document Prepared for the Reviewers of the Center for Integrated Agricultural Systems. University of Wisconsin.

Cochrane, Willard. 1979. *The Development of American Agriculture: A Historical Analysis*. University of Minnesota Press.

Colborn, Theo, Dianne Dumanoski, and John Peterson Myers. 1996. *Our Stolen Future: Are We Threatening Our Fertility, Intelligence, and Survival? A Scientific Detective Story*. Dutton.

Conaway, James. 1990. *Napa*. Houghton Mifflin.

Conaway, James. 2002. *The Far Side of Eden*. Houghton Mifflin.

Costello, Michael, and Kent Daane. 1998. Arthropods. In *Cover Cropping in Vineyards*, ed. C. Ingels, R. Bugg, G. McGourty, and P. Christensen. University of California Division of Agriculture and Natural Resources.

Council for Agricultural Science and Technology. 1990. Alternative Agriculture: Scientists' Review.

Crovetto Lamarca, Carlos. 1996. *Stubble over the Soil*. American Society of Agronomy.

Culver, Dennis. 1993. Lodi-Woodbridge District-Wide IPM program. In *The Greening of the California Grape and Wine Industry*, ed. M. Thompson. Senate Committee on California's Wine Industry.

Daane, Kent, and Michael Costello. 1998. Can Cover Crops Reduce Leafhopper Abundance in Vineyards? *California Agriculture* 52, no. 5: 27–33.

Division of Agricultural Sciences. 1978. *Pear Pest Management*. University of California.

Dlott, Jeff. 2004. California Wine Community Sustainability Report: Executive Summary.

Dlott, Jeff, Miguel Altieri, and Mas Masumoto. 1994. Exploring the Theory and Practice of Participatory Research in US Sustainable Agriculture: A Case Study in Insect Pest Management. *Agriculture and Human Values* 11, no. 2: 126–139.

Dlott, Jeff, Thomas Nelson, Robert Bugg, Mike Spezia, Ray Eck, Judith Redmond, and Liza Lewis. 1996. California, USA: Merced County BIOS project. In *New Partnerships for Sustainable Agriculture*, ed. L. Thrupp. World Resources Institute.

Donovan, Peter. 1999. Wilke Team Designs a No-Till Future. *Patterns of Choice* 9: 1–10.

Dunlap, Thomas R. 1981. *DDT: Scientists, Citizens, and Public Policy*. Princeton University Press.

Dunley, John, and Stephen Welter. 2000. Correlated Insecticide Cross-Resistance in Azinphosmethyl Resistant Codling Moth (Lepidoptera: Tortricidae). *Journal of Economic Entomology* 93, no. 3: 955–962.

Elliott, Bob, Larry Wilhoit, Madeline Brattesani, and Nan Gorder. 2004. Pest Management Assessment for Almonds Reduced-Risk Alternatives to Dormant Organophosphate Insecticides. California Department of Pesticide Regulation.

Elliott, E. T., and C. V. Cole. 1989. A Perspective on Agroecosystem Science. *Ecology* 70, no. 6: 1597–1602.

El Titi, Adel. 1992. Integrated Farming: An Ecological Farming Approach in European Agriculture. *Outlook on Agriculture* 21, no. 1: 33–39.

Epstein, Lynn, and Susan Bassein. 2003. Patterns in Pesticide Use in California and The Implications for Strategies for Reductions of Pesticides. *Annual Review of Phytopathology* 41: 23.1–23.25.

Epstein, Lynn, Susan Bassein, and Frank Zalom. 2000. Almond and Stone Fruit Growers Reduce OP, Increase Pyrethroid Use in Dormant Sprays. *California Agriculture* 54, no. 6: 14–19.

Epstein, Lynn, Susan Bassein, Frank Zalom, and Larry Wilhoit. 2001. Changes in Pest Management Practice in Almond Orchards During the Rainy Season in California, USA. *Agriculture, Ecosystems, and Environment* 83: 111–120.

Farmers for Alternative Agricultural Research. 1990. Reducing the Use of Pesticides in Agriculture: A Farmers' Perspective. California Action Network, California Association of Family Farmers, and Committee for Sustainable Agriculture.

Feise, Chris, and Nicholas Lovrich. 2003. Reflections on the EPA Columbia Plateau Agricultural Initiative (CPAI) by Program Participants: Outcomes Associated with a Community Based Approach to Environmental Protection. University of Washington Division of Governmental Studies and Services.

Fisk, John, Oran Hesterman, and Thomas Thorburn. 1998. Integrated Farming Systems: A Sustainable Agriculture Learning Community in the USA. In *Facilitating Sustainable Agriculture*, ed. N. Rolling and M. Wagemakers. Cambridge University Press.

Fiske, Emmett. 1979. The College and Its Constituency: Rural and Community Development at the University of California 1875–1978, Social and Economic Development. University of California, Davis.

FitzSimmons, Margaret, and David Goodman. 1998. Incorporating Nature: Environmental Narratives and the Reproduction of Food. In *Remaking Reality*, ed. B. Braun and N. Castree. Routledge.

Foe, Chris. 1987. Orchard and Alfalfa Toxicity Study. Central Valley Regional Water Quality Control Board.

Friedland, William. 1980. Technology in Agriculture: Labor and the Rate of Accumulation. In *The Rural Sociology of the Advanced Societies*, ed. F. Buttel and H. Newby. Allanheld.

Friedland, William. 1984. Commodity Systems Analysis: An Approach to the Sociology of Agriculture. In *Research in Rural Sociology and Development*, ed. H. Schwarzweller. JAI.

Friedland, William. 2001. Reprise on Commodity Systems Methodology. *International Journal of Sociology of Agriculture and Food* 9, no. 1: 82–103.

Friedland, William. 2002. Agriculture and Rurality: Beginning the "Final Separation"? *Rural Sociology* 67, no. 3: 350–371.

Friedland, William, and Amy Barton. 1975. Destalking the Wily Tomato: A Case Study in Social Consequences in California Agricultural Research. Department of Applied Behavioral Sciences, University of California, Davis.

Friedland, William, and Alan Haight. 1985. Marketing Orders: A Sociological Perspective. Unpublished.

Friedland, William, Amy Barton, and Robert Thomas. 1981. *Manufacturing Green Gold: Capital, Labor and Technology in the Lettuce Industry*. Cambridge University Press.

Fujimoto, Isao, and Emmett Fiske. 1975. What Research Gets Done at a Land Grant College: Internal Factors at Work. Paper read at meeting of Rural Sociological Society, San Francisco.

Fujimoto, Isao, and William Kopper. 1975. Outside Influences on What Research Gets Done at a Land Grant School: Impact of Marketing Orders. Paper read at meeting of Rural Sociological Society, San Francisco.

Gibbs, Marcia, Rex Dufour, and Martin Guerena. 2005. BASIC Cotton Manual. Sustainable Cotton Project.

Glades Crop Care Inc. 2003. Florida Fruit & Vegetable IPM Innovations Tour.

Gliessman, Stephen. 1998. *Agroecology: Ecological Processes in Sustainable Agriculture*. Ann Arbor Press.

Glotfelty, D. E., J. N. Seiber, and L. A. Liljedahl. 1987. Pesticides in Fog. *Nature* 325, February 12: 602–605.

Glover, Jerry. 2005. The Necessity and Possibility of Perennial Grain Production Systems. *Renewable Agriculture and Food Systems* 20, no. 1: 1–4.

Gordon, Rosemary. 2002. A Recipe for Success. *American Vegetable Grower*, September.

Gottlieb, Robert. 2001. *Environmentalism Unbound: Exploring New Pathways for Change*. MIT Press.

Groth, Edward, Charles Benbrook, Karen Benbrook, and Adam Goldberg. 2001. A Report Card for the EPA: Successes and Failures in Implementing the Food Quality Protection Act. Consumers Union.

Guthman, Julie. 1998. Regulating Meaning, Appropriating Nature: The Codification of California Organic Agriculture. *Antipode* 30, no. 2: 135–154.

Guthman, Julie. 2000. Raising Organic: Grower Practices in California. *Agriculture and Human Values* 17: 257–266.

Guthman, Julie. 2004a. Back to the Land: The Paradox of Organic Food Standards. *Environment and Planning* A 36: 511–528.

Guthman, Julie. 2004b. *Agrarian Dreams: The Paradox of Organic Farming in California*. University of California Press.

Harrison, Jill. 2004. Invisible People, Invisible Places: Connecting Air Pollution and Pesticide Drift in California. In *Smoke and Mirrors*, ed. E. Dupuis. NYU Press.

Hassanein, Neva. 1999. *Changing the Way America Farms: Knowledge and Community in the Sustainable Agriculture Movement*. University of Nebraska Press.

Hassanein, Neva, and Jack Kloppenburg. 1995. Where the Grass Grows Again: Knowledge Exchange in the Sustainable Agriculture Movement. *Rural Sociology* 60, no. 4: 721–740.

Hayes, Tyrone. 2005. Welcome to the Revolution: Integrative Biology and Assessing the Impact of Endocrine Disruptors on Environmental and Public Health. *Integrative and Comparative Biology* 45: 321–329.

Henderson, George. 1999. *California and the Fictions of Capital*. Oxford University Press.

Hendrickson, Mary, and William Heffernan. 2002. Opening Spaces through

Relocalization: Locating Potential Resistance in the Weaknesses of the Global Food System. *Sociologia Ruralis* 42, no. 4: 347–369.

Hightower, Jim. 1973. *Hard Tomatoes, Hard Times*. Schenkman.

Holt-Giménez, Eric. 2002. Measuring Farmers' Agroecological Resistance after Hurricane Mitch in Nicaragua: A Case Study in Participatory, Sustainable Land Management Impact Monitoring. *Agriculture, Ecosystems and Environment* 93: 87–105.

Howarth, Robert, Donald Anderson, James Cloem, Chris Elfring, Charles Hopkinson, Brian Lapointe, Tom Malone, Nancy Marcus, Karen McGlathery, Andrew Sharpley, and Dan Walker. 2000. Nutrient Pollution of Coastal Rivers, Bays and Seas. *Issues in Ecology* 7.

Hynes, H. Patricia. 1989. *The Recurring Silent Spring*. Pergamon.

Ingels, Chuck, and Karen Klonsky. 1998. Historical and Current Uses. In *Cover Cropping in Vineyards*, ed. C. Ingels, R. Bugg, G. McGourty, and P. Christensen. University of California Division of Agriculture and Natural Resources.

Ingels, Chuck, Robert Bugg, Glenn McGourty, and Peter Christensen. 1998. Cover Cropping in Vineyards: A Grower's Handbook. University of California Division of Agriculture and Natural Resources.

Kimbrell, Andrew, ed. 2002. *Fatal Harvest: The Tragedy of Industrial Agriculture*. Island.

Kingdon, John. 1995. *Agendas, Alternatives and Public Policies*, second edition. Addison-Wesley.

Klonsky, Karen, Frank Zalom, and W. Barnett. 1990. California's Almond IPM Program. *California Agriculture* 44, no. 5: 21–24.

Kloppenburg, Jack. 1988. *First The Seed: The Political Economy of Plant Biotechnology*. Cambridge University Press.

Kloppenburg, Jack. 1991. Social Theory and the De/reconstruction of Agricultural Science: Local Knowledge for an Alternative Agriculture. *Rural Sociology* 56, no. 4: 519–548.

Kuivila, K., and C. Foe. 1995. Concentration, Transport, and Biological Impact of Dormant Spray Pesticides in the San Francisco Estuary, California. *Environmental Toxicology and Chemistry* 14: 1141–1150.

Lacy, William. 2000. Commercialization of university research brings benefits, raises issues and concerns. *California Agriculture* 54, no. 4: 72–79.

Land Institute. 2003. The Land Report.

Lapsley, James. 1996. *Bottled Poetry: Napa Winemaking from the Prohibition to the Modern Era*. University of California Press.

Larson, Magali Sarfatti. 1977. *The Rise of Professionalism: A Sociologial Analysis*. University of California Press.

Latour, Bruno. 1987. *Science in Action: How to Follow Scientists and Engineers Through Society*. Harvard University Press.

Latour, Bruno. 1999. On Recalling ANT. In *Actor Network Theory and After*, ed. J. Law and J. Hassard. Blackwell.

Learn, Elmer, James Lyons, and James Meyers. 1987. Strategic Planning Phase I. Manuscript, University of California Division of Agriculture and Natural Resources.

Legner, E. Fred, and Gordon Gordh. 1992. Lower Navel Orangeworm (Lepidopterae: Phycitidae) Population Densities Following Establishment of *Goniozus legneri* (Hymenoptera: Bethylidae) in California. *Journal of Economic Entomology* 85, no. 6: 2153–2160.

Lighthall, David. 1995. Farm Structure and Chemical Use. *Rural Sociology* 60, no. 3: 505–520.

Lighthall, David. 1996. Sustainable Agriculture in the Corn Belt: Production Side Progress and Demand Side Constraints. *American Journal of Alternative Agriculture* 11, no. 4: 168–174.

Lighthall, David, and Rebecca Roberts. 1995. Towards an alternative logic of technological change: Insights from Corn Belt agriculture. *Journal of Rural Studies* 11, no. 3: 319–334.

Long, N., and A. Long, eds. 1992. *Battlefields of Knowledge*. Routledge.

Lucido, Frank. 1999. Dryland Wheat Growers and the EPA Region X, Community-Based Columbia Plateau Agricultural Initiative. Seattle: US/EPA, Region X.

Lynch, Sarah, Deana Sexson, Chuck Benbrook, Mike Carter, Jeff Wyman, Pete Nowak, Jeb Barzen, Steve Diercks, and John Wallendal. 2000. Accelerating Industry-Wide Transition to Reduced-Risk Pest Management Systems: A Case Study of the Wisconsin Potato Industry. *Choices*, September.

Lyons, James. 2004. *A History of the UC Statewide IPM Program*. University of California Statewide IPM Program.

Marshall, Andrew. 2000. Sustaining Sustainable Agriculture: The Rise and Fall of the Fund for Rural America. *Agriculture and Human Values* 17: 267–277.

Marthedal, Jon. 2001. Sun-Maid Integrated Pest Management/Best Management Practices Program. In *Partnerships for Sustaining California Agriculture*, ed. D. Cheney, L. Halprin and E. Dabbs. University of California Division of Agriculture and Natural Resources/SAREP.

Masumoto, Mas. 1995. *Epitaph for a Peach: Four Seasons on My Family Farm*. Harper.

McConnell, Grant. 1959. *Decline of Agrarian Democracy*. University of California Press.

McDavit, W. Michael, and Kris O'Connor. 2004. Expanding Environmental Policy Networks: The Central Coast Vineyard Team and the Pesticide Environmental Stewardship Program. Presented at meeting of American Society for Public Administration, Portland, Oregon.

McDowell, George. 2001. *Land-Grant Universities and Extension into the 21st Century: Re-Negotiating or Abandoning a Social Contract*. Iowa State University Press.

McNair, Joel. 1992a. Grandpa Never Had It This Good. *Agri-view*, February: 1.

McNair, Joel. 1992b. The "New" Grazing: Where Dairy Heresy Is Spoken. *Agri-view* 18, March 5: 1.

Meffe, Gary, and C. Ronald Carroll. 1994. *Principles of Conservation Biology.* Sinauer Associates.

Metcalfe, Mark, Bruce McWilliams, Brent Hueth, Robert Van Steenwyk, David Sunding, Aaron Swoboda, and David Zilberman. 2002. The Economic Importance of Organophosphates in California Agriculture. California Department of Food and Agriculture.

Mitchell, Jeffrey. 2001. Innovative Agricultural Extension Partnerships in California's Central San Joaquin Valley. *Journal of Extension* 39, no. 6: 1–12.

Mitchell, Jeffrey, Tim Hartz, Stu Pettygrove, Daniel Munk, Donald May, Frank Menezes, John Diener, and Tim O'Neill. 1999. Organic Matter Recycling Varies with Crops Grown. *California Agriculture* 53, no. 4: 37–40.

Mondavi, Robert. 1998. *Harvests of Joy: My Passion for Excellence.* Harcourt Brace.

Morris, Carol. 2000. A "Quiet Revolution"? Integrated Farming Systems in the Transition to a More Sustainable Agriculture in the U.K. In *Agricultural and Environmental Sustainability in the New Countryside*, ed. H. Milward, K. Beesley, B. Ilbery and L. Harrington. St. Mary's University and Nova Scotia Agricultural College.

Morris, Carol, and Michael Winter. 1999. Integrated Farming Systems: The Third 408for European Agriculture? *Land Use Policy* 16: 193–205.

Morse, Steve. 2004. Green Lands, Blue Waters Project Description Document.

Murphy, Broo, Jay Rosenheim, and Jeffrey Grannett. 1996. Habitat Diversification for Improving Biological Control: Abundance of *Anagrus epos* (Hymenoptera: Mymaridae) in Grape Vineyards. *Environmental Entomology* 25, no. 2: 495–504.

Napa Sustainable Winegrowing Group. 1997. Integrated Pest Management Field Handbook for Napa County.

National Research Council. 1989. *Alternative Agriculture.* National Academy Press.

National Research Council. 1993. *Soil and Water Quality: An Agenda for Agriculture.* National Academy Press.

National Research Council. 1995. *Colleges of Agriculture at the Land Grant Universities: A Profile.* National Academy Press.

National Research Council. 1996. *Colleges of Agriculture at the Land Grant Universities: Public Service and Public Policy.* National Academy Press.

National Research Council. 2000. *Clean Coastal Waters: Understanding and Reducing the Effects of Nutrient Pollution.* National Academy Press.

National Resources Defense Council. 1998. Fields of Change: A New Crop of American Farmers Finds Alternatives to Pesticides.

Ohmart, Clifford. 1998. Lodi-Woodbridge Winegrape Commission's

Biologically Integrated Farming Systems for Winegrapes, Final Report. Lodi-Woodbridge Winegrape Commission.

Ohmart, Clifford. 2001. Lodi-Woodbridge Winegrape Commission's Sustainable Farming Program: The Role of Education, Outreach and Marketing. In *Finding The Right Blend*, ed. J. Lapsley and J. Loux. University of California Davis University Extension.

Ohmart, Clifford. 2004. Twelve Years of Sustainable Viticulture in Lodi. *LWWC Research-IPM Newsletter*, July: 1–3.

Ohmart, Clifford, and Stephen Matthiasson. 2000. Lodi-Woodbridge Winegrowers Workbook. Lodi-Woodbridge Winegrape Commission.

Olson, William, Carolyn Pickel, Richard Buchner, William Krueger, Franz Niederholzer, Max Norton, Wilbur Reil, Steve Sibbett, Fred Thomas, and Larry Whitted. 2003. Integrated Prune Farming Practices Decision Guide. University of California Division of Agriculture and Natural Resources.

Osmond, Deanna. 2002. Final Report Neuse Crop Management Project. Soil Science Department, North Carolina State University.

Pence, Robert. 1998. Leveling the Learning Fields: An Assessment of the Agriculture Partnership of BIOS-Merced and BIFS-Lodi. Department of Human and Community Development, University of California.

Pence, Robert, and James Grieshop. 2001. Mapping the Road for Voluntary Change: Partnerships in Agricultural Extension. *Agriculture and Human Values* 18: 209–217.

Peters, Scott. 1996. *Cooperative Extension and the Democratic Promise of the Land Grant Idea*. Minnesota Extension Service.

Pew Oceans Commission. 2003. America's Living Oceans: Charting a Course for Sea Change.

Pickel, Carolyn, Walt Bentley, Joseph Connell, Roger Duncan, and Mario Viveros. 2004. A Seasonal Guide to Environmentally Responsible Pest Management in Almonds. University of California Division of Agriculture and Natural Resources.

Pincetl, Stephanie. 1999. *Transforming California: A Political History of Land Use and Development*. Johns Hopkins University Press.

Pretty, Jules. 1998. Supportive Policies and Practice for Scaling Up Sustainable Agriculture. In *Facilitating Sustainable Agriculture*, ed. N. Rolling and M. Wagemakers. Cambridge University Press.

Ransom, Bev, Joseph Grant, and Janet Broome. 2003. Walnut BIFS Project: San Joaquin County Walnut Growers Survey. *Sustainable Agriculture* 14, no. 3: 3–5.

Ravetz, Jerome. 1988. Ethics in Scientific Activity. In *Professional Ideals*, ed. A. Flores. Wadsorth.

Roberts, Diana. 2001. The Wilke Project: A Grower- and Community-Driven Project for Direct Seeding in Eastern Washington. Paper read at Partnerships for Sustaining California Agriculture conference, Woodland, California.

Rogers, Everett. 1983. *Diffusion of Innovations*, third edition. Free Press.

Röling, Niels. 1988. *Extension Science: Information Systems In Agricultural Development*. Cambridge University Press.

Röling, Neils, and Annemarie Wagemakers. 1998. *Facilitating Sustainable Agriculture: Participatory Learning and Adaptive Management in Times of Environmental Uncertainty*. Cambridge University Press.

Rosenbaum, Walter. 1994. The Clenched Fist and the Open Hand: Into the 1990s at EPA. In *Environmental Policies in the 1990s*, ed. N. Vig and M. Kraft. Congressional Quarterly Inc.

Rosenberg, Charles. 1961. *No Other Gods: On Science and American Social Thought*. Johns Hopkins University Press.

Ruttan, Vernon. 1996. What Ever Happened to Adoption-Diffusion Research? *Sociologia Ruralis* 36, no. 1: 51–73.

Ryan, Bryce, and Neal Gross. 1943. The Diffusion of Hybrid Seed Corn in Two Iowa Communities. *Rural Sociology* 8: 15–23.

Saker, Victoria Alice. 1990. Benevolent Monopoly: The Legal Transformation of Agricultural Cooperation, 1890–1943. Dissertation, University of California, Berkeley.

SAREP (Sustainable Agriculture Research and Education Program). 1998. *Eco Labeling for California Winegrapes*, ed. J. Broome, C. Ohmart, A. Moskow, and J. Waddle.

SAREP. 1999. Request for Proposals. In Biologically Integrated Farming Systems (BIFS) Program: A Progress Report to the California State Legislature on the Implementation of Assembly Bill 3383. University of California Division of Agriculture and Natural Resources.

Savory, Allan. 1988. *Holistic Resource Management*. Island.

Schaffer, Kristin, editor. 1997. Learning from the BIOS Approach: A Guide for Community-Based Biological Farming Programs. Community Alliance with Family Farmers.

Schlosser, Eric. 2001. *Fast Food Nation: The Dark Side of the All-American Meal*. Houghton Mifflin.

Scott, Roy. 1970. *The Reluctant Farmer: The Rise of Agricultural Extension to 1914*. University of Illinois Press.

Sexson, D., and J. Wyman. 2005. Movement of Colorado Potato Beetle (Coleoptera: Chrysomelidae) between Commercial Potato Fields in Wisconsin: Development of Area-Wide Colorado Potato Beetle Pest Management Strategies. *Journal of Econonomic Entomolology* 98, no. 3: 716–724.

Sharp, Gwen, Chris Carusi, Chuck Francis, Heidi Moullessaux-Kunzman, Susan Smalley, Steve Stevenson, and Sean Swezey. 2000. Collaborating for Change: Proceedings of a Gathering of Sustainable Agriculture Programs. University of Wisconsin Center for Integrated Agricultural Systems.

Smart, James. 1998. All the Stars in the Heavens Were in the Right Places: The

Passage of the Food Quality Protection Act of 1996. *Stanford Environmental Law Journal* 17, no. 2: 273–352.

Sonoma County Grape Growers Association. 2003a. IPM Fieldbook.

Sonoma County Grape Growers Association. 2003b. Decreasing Pesticide Usage in Sonoma County. Press release.

Steggall, John. 2003. Assessing the Importance of Organophosphates in California Agriculture. Presented to PUR Workgroup, October 24.

Stevenson, George, Joshua Posner, John Hall, Lee Cunningham, and Jan Harrison. 1994. Addressing the Challenges of Sustainable Agriculture Research and Extension at Land-Grant Universities: Radially Organized Teams at Wisconsin. *American Journal of Alternative Agriculture* 9, no. 1: 76–83.

Rockefeller Foundation. 1982. Science for Agriculture: A Report on Critical Issues in American Agricultural Research.

US Geological Survey. 1998a. Water Quality in the San Joaquin-Tulare Basins: California 1992–1995. Circular 1159.

US Geological Survey. 1998b. Water Quality in the Central Columbia Plateau: Washington and Idaho 1992–1995. Circular 1144.

US Geological Survey. 1999. The Quality of Our Nation's Waters: Nutrients and Pesticides. Circular 1225.

US Geological Survey. 2000. Water Quality in the Sacramento River Basin, California, 1994–1998.

UC-IPM (University of California Statewide Integrated Pest Management Project). 2002. Integrated Pest Management for Almonds, second edition. Division of Agriculture and Natural Resources.

Uphoff, Norman, ed. 2002. *Agroecological Innovations: Increasing Food Production with Participatory Development*. Earthscan.

USEPA (US Environmental Protection Agency). 1997a. Community Based Environmental Protection: A Resource Book for Protecting Ecosystems and Communities.

USEPA. 1997b. Community Based Environmental Protection Strategy, Region X.

USEPA. 1997c. Proposal for FY97: The Columbia Plateau Agricultural Initiative.

USEPA. 1998. Columbia Plateau Agricultural Initiative. Environmental Fact Sheet 6.

USEPA. 2000a. Diazinon Technical Briefing.

USEPA. 2000b. Overview of Diazinon: Revised Risk Assessment.

USEPA. 2004. Food Quality Protection Act (FQPA) of 1996.

Van den Bosch, Robert. 1978. *The Pesticide Conspiracy*. University of California Press.

Vos, Timothy. 2000. Visions of the Middle Landscape: Organic Farming and the Politics of Nature. *Agriculture and Human Values* 17: 245–256.

Walker, Richard. 2004. *The Conquest of Bread: 150 Years of California Agribusiness*. New Press.

Warner, Keith Douglass. 2004. Agroecology in Action: How the Science of Alternative Agriculture Circulates through Social Networks. Dissertation, University of California, Santa Cruz.

Warner, Keith Douglass. 2006a. Extending Agroecology: Grower Participation in Partnerships Is Key to Social learning. *Renewable Food and Agriculture Systems* 21: 84–94.

Warner, Keith Douglass. In review (b). Quality In Place: Agroecological Partnerships and the Geographic Branding of California Winegrapes. *Journal of Rural Studies.*

Warner, Keith Douglass. In review (c). Agroecology as Participatory Science: Emerging Alternatives to Technology Transfer Extension Practice. *Science, Technology & Human Values.*

Wasserman, Stanley, and Katherine Faust. 1994. *Social Network Analysis: Methods and Applications.* Cambridge University Press.

Welter, Stephen, Carolyn Pickel, Jocelyn Millar, Frances Cave, Robert Van Steenwyk, and John Dunley. 2005. Pheromone Mating Disruption Offers Selective Management Options for Key Pests. *California Agriculture* 59, no. 1: 16–22.

Whatmore, Sarah, and Lorraine Thorne. 1997. Nourishing Networks: Alternative Geographies of Food. In *Globalising Food*, ed. D. Goodman and M. Watts. Routledge.

Wondolleck, Juli, and Steven Yaffee. 2000. *Making Collaborations Work: Lessons from Innovation in Natural Resource Management.* Island.

Woodhill, James, and Niels Röling. 1998. The Second Wing of the Eagle: The Human Dimension in Learning Our Way to More Sustainable Futures. In *Facilitating Sustainable Agriculture*, ed. N. Röling and M. Wagemakers. Cambridge University Press.

Woodworth, Charles. 1912. The Insecticide Industries in California. *Journal of Economic Entomology 5*, August: 358.

Wright, Angus. 1990. *The Death of Ramon Gonzalez: The Modern Agricultural Dilemma.* University of Texas Press.

Zhang, Minghua, Larry Wilhoit, and Chris Geiger. 2004. Dormant Season Organophosphate Use in California Almonds. California Department of Pesticide Regulation.

Index

Adoption, 44

Agricultural policy, 22, 23, 89, 92, 93, 99, 223–226, 230–232

Agroecological partnerships, 5, 17, 67–87, 194, 211, 224, 226

Agroecology, 3, 25–29, 87, 135–137, 223, 224, 226

Almond Board of California, 17, 19, 69, 70, 120–122, 131, 177–181, 212–214

Almonds, 14–17, 59–70, 83, 84, 125–135, 176–181, 212–214

Alternative Agriculture (National Research Council report), 41, 42, 55, 56, 60, 230, 231

Anderson, Glenn, 14, 15, 59–62, 85, 86, 105, 106, 132–134

Annual Cropping, Intense Rotation, Direct Seed (ACIRDS), 94, 95, 98–100, 111, 117, 119, 152

Azinphosmethyl. *See* Organophosphates

Benbrook, Chuck, 218, 219

Biological Control, 14, 46, 47, 61, 131–134, 142, 146–148, 170

Biologically Integrated Farming Systems (BIFS), 17, 70–73, 76, 145, 151, 152, 155, 167, 182–184, 188, 203–209. *See also* Sustainable Agriculture Research and Education Program

Biologically Integrated Orchard Systems (BIOS), 16–19, 59–72, 78, 79, 86, 87, 106, 133–135, 138, 144, 152–154, 159, 176, 177

Bugg, Bob, 16, 60–64, 70, 106, 149

California Association of Winegrape Growers, 192, 210, 211

California Department of Pesticide Regulation, 69, 73–75, 122–124, 171, 192, 225

Carson, Rachel, 1, 2, 13, 15, 21–24, 32–34, 42, 47, 108, 232

Center for Agricultural Partnerships, 77, 172

Center for Integrated Agricultural Systems (CIAS), 42, 55, 173

Central Coast Vineyard Team, 155, 186, 190–192

Clean Water Act (Federal Water Pollution Control Act Amendments), 22, 25, 123

Code of Sustainable Winegrowing, 20, 156, 193, 211

Codling moth, 9, 14, 138, 143, 144, 163–173

Columbia Plateau Agricultural Initiative (CPAI), 95–99, 123, 124

Commodity organization, 74, 119–122, 225

Community Alliance with Family Farms (CAFF), 16, 17, 60–63, 66–72, 78, 106, 121, 122, 134, 135, 176–181, 228

Conservation tillage, 90–92, 151
Cooperative Extension. *See*
    Extension; University of California
    Cooperative Extension Service
Cover crops, 61, 148–153
Crop subsidies. *See* Agricultural
    policy

Diazinon. *See* Organophosphates
Diffusion, 44
Direct Seed. *See* Conservation tillage;
    Annual Cropping, Intense Rotation,
    Direct Seed
Dlott, Jeff, 61, 76, 156, 220

Ecolabel, 20, 205, 213, 214
Environmental policy, 22–25, 227–232
Environmental Protection Agency, US,
    2, 22–25, 65, 66, 70, 73–77, 95–99,
    121–124, 191, 223, 225–227, 232
Extension, 3, 4, 15, 42, 45, 51–53,
    60–63, 66, 67, 77, 81, 82,
    110–117, 131, 134, 135, 153, 159,
    173, 229

Feder, Augie, 65, 66, 69, 213
Food Quality Protection Act, 25, 75,
    76, 83, 217
Franciscans, 125, 197

Gerber, 171, 172, 214, 215
Green Lands, Blue Waters, 226, 227
Growers
    enrolled, 102
    farm-management styles of, 104–108
    leading, 101, 102
    of perennial crops, 103, 104
    progressive, 103, 107
Guthion. *See* Organophosphates

Heintz, Chris, 121, 177, 178, 181,
    214
Hemly, Doug, 9–14, 101, 137, 166,
    170, 181, 182
Hendricks, Lonnie, 15, 16, 59–65, 82,
    132–134

Integrated farming systems, 56, 72,
    77, 139, 152–156, 214
Integrated Pest Management (IPM), 9,
    11–14, 19, 25, 46–48, 59, 61, 66,
    131, 132, 143, 164–170, 202, 203,
    214, 217–220
Intentional rotational grazing, 35–39,
    173
Iowa State University, 40, 41, 228

Kellogg Foundation, 77, 203
Kupers, Karl, 89–94, 98–101, 111

Land-grant universities, 42–57, 108,
    227–230
Lange, Randy, 18, 101, 106, 200,
    207, 208
Latour, Bruno, 29–32, 57, 87, 99,
    124, 136, 160, 173, 201, 202, 209,
    215, 225, 227
Ledbetter, John, 18, 19, 101, 207
Leopold Center, 41, 54, 227
Lodi winegrape partnership, 18–20,
    155, 156, 200–209, 223
Looker, Mark, 121, 177–179

Mills, Nicholas, 145
Monitoring, 139–145, 169, 170
Monoculture, 17, 36, 92, 93, 157,
    159, 224

Napa Sustainable Winegrowing
    Group, 186–188, 210
Navel orange worm, 126–135
Nutrient pollution, 23, 24, 36, 148,
    222, 227

Ohmart, Cliff, 19, 155, 156, 203–207
Orchard sanitation, 130, 142, 168
Organic agriculture, 14–16, 29, 153
Organophosphates, 11–14, 17, 65,
    66, 75, 76, 83–85, 130–132, 139,
    142, 146–148, 165, 169, 183

Pears, 9, 12–14, 77, 84, 85, 137–139,
    164–172, 181, 182, 214, 215, 229

Pest control advisors (PCAs), 9, 59, 64, 79, 110–115, 131, 132, 142–145
Pest Management Alliance (PMA), 70, 73–75, 86, 117, 122, 165–171, 177–181, 193, 212
Pesticide Use Reporting (PUR), 73, 80–83
Pheromone mating disruption, 11–14, 77, 82, 84, 137, 138, 143, 144, 147, 163–172, 194
Potatoes, 217–223
Practical Farmers of Iowa, 40, 55, 230
Productionist ideology, 45, 49
Protected Harvest, 205, 220–223, 226, 232
Prunes, 65, 66, 69, 154, 182–184

Randall Island Project, 12–14, 85, 137, 138, 168–170, 181, 214
Reed, Rick, 16, 60–65, 70
Ridge-till farming, 40. *See also* Conservation tillage
Risk, 62, 112–114, 156–159

Science and Technology Studies (STS), 29–34, 108, 173, 226
Silent Spring. See Carson, Rachel
Social learning, 1, 3, 4, 16, 72, 81, 85, 99, 124, 130, 131, 135–144, 169, 201, 224–226
Sonoma County Grape Growers Association, 122, 188–190,
Sun-Maid, 76, 122, 215
Sustainability, 20, 28, 205, 207, 210, 211, 231
Sustainable agriculture, 4, 28, 29, 50, 53–55, 228–230
Sustainable Agriculture Research and Education Program (SAREP), 15–19, 55, 60, 70–72, 132, 228. *See also* Biologically Integrated Farming Systems

Transfer of technology, 3, 52, 81, 82, 136, 173, 224

University of California, 45–49, 108–110, 115–119, 143, 228, 229
University of California Cooperative Extension Service (UCCE), 15, 53, 62–67, 81, 82, 85, 86, 108–110, 115–117, 134, 135, 159, 177–184
University of California Integrated Pest Management Program, 47, 48, 116, 117, 131, 147, 177, 178, 229
University of Wisconsin, 35–39, 219–221, 228. *See also* Center for Integrated Agricultural Systems

Walnuts, 77, 82, 138, 139, 164–167
Washington State University, 93–95
Weddle, Pat, 9, 11, 13, 19, 26, 84, 154, 165–170, 182, 194
Welter, Stephen, 11–13, 19, 103, 118, 168, 183
Wilke Farm, 94, 95, 98, 99
Winegrapes, 17–20, 85, 138, 139, 155, 156, 184–193, 197–211, 229
Wine Institute, 156, 211
Wisconsin Potato and Vegetable Growers Association, 217–223
World Wildlife Fund, 217–223